The independent airport planning manual

Related titles:

Reference calculations for planning airport terminal buildings: a supplement to ***The independent airport planning manual***
(ISBN 978-0-85709-140-6)
This CD-Rom contains calculations which enable the core passenger processing and retail components of the internal space within a terminal building to be calculated statically. These calculations will be an invaluable support to terminal planners and architects developing the masterplan size of a terminal building.

The air transport system
(ISBN 978-1-84569-325-1)
'. . . an excellent and well-written book.'

The Aerospace Professional

This important book reviews the major operational issues in planning and managing air transport from the point of view of airports, airlines and manufacturers.

Aircraft system safety: military and civil aeronautical applications
(ISBN 978-1-84569-136-3)
'A must-have text for all serious practitioners of system safety.'
Terry J. Gooch, Embry Riddle Aeronautical University

This authoritative text reviews safety concepts, hazards, safety assessment and management in aircraft electrical systems.

Details of these and other Woodhead Publishing books can be obtained by:

- visiting our web site at www.woodheadpublishing.com
- contacting Customer Services (e-mail: sales@woodheadpublishing.com; fax: +44 (0) 1223 893694; tel.: +44 (0) 1223 891358 ext. 130; address: Woodhead Publishing Limited, Abington Hall, Granta Park, Great Abington, Cambridge CB21 6AH, UK)

If you would like to receive information on forthcoming titles, please send your address details to: Francis Dodds (address, tel. and fax as above; e-mail: francis.dodds@woodheadpublishing.com). Please confirm which subject areas you are interested in.

The independent airport planning manual

Alexandre L. W. Bradley

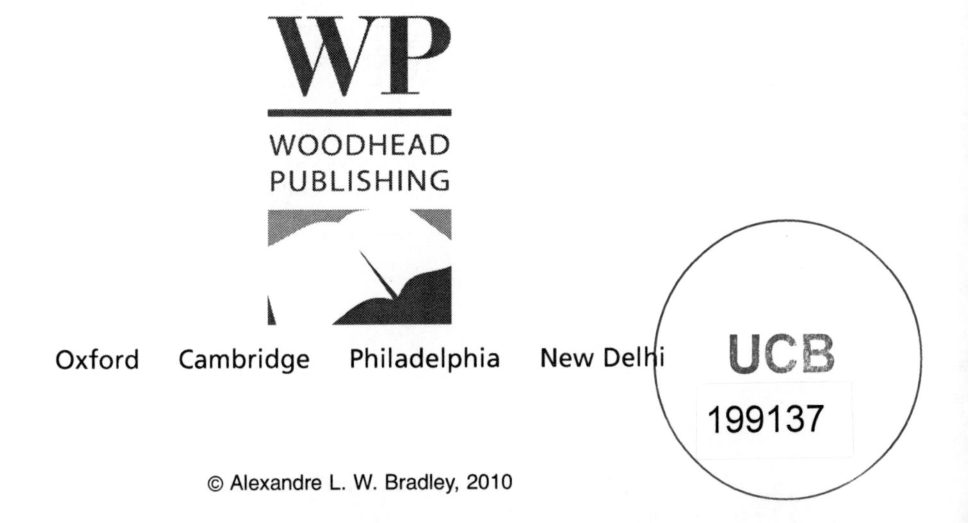

Published by Woodhead Publishing Limited, Abington Hall, Granta Park, Great Abington, Cambridge CB21 6AH, UK
www.woodheadpublishing.com

Woodhead Publishing, 525 South 4th Street #241, Philadelphia, PA 19147, USA

Woodhead Publishing India Private Limited, G-2, Vardaan House, 7/28 Ansari Road, Daryaganj, New Delhi – 110002, India
www.woodheadpublishingindia.com

First published 2010, Woodhead Publishing Limited

British Library Cataloguing in Publication Data
A catalogue record for this book is available from the British Library.

Woodhead Publishing ISBN 978-1-84569-713-6 (print)
Woodhead Publishing ISBN 978-0-85709-035-5 (online)

The publisher's policy is to use permanent paper from mills that operate a sustainable forestry policy, and which has been manufactured from pulp which is processed using acid-free and elemental chlorine-free practices. Furthermore, the publisher ensures that the text paper and cover board used have met acceptable environmental accreditation standards.

Typeset by Data Standards Limited, Frome, Somerset, UK

Cover image supplied courtesy of Edmund Sumner Photography – Architects: Grimshaw.

This book is dedicated to my wife Helen Bradley and my late father Leonard Bradley.

Contents

About the author

Alexandre Bradley BEng (Hons), CEng, MIMechE, MPhil has been an employee of BAA (formerly British Airports Authority) for 16 years in total. During this time he worked as a Senior Mechanical Engineer, Stansted Generation 2 Terminal Planning Manager, and then as Head of Terminal and Satellites for Stansted Generation 1 major devlopments. Alex Bradley is a Chartered Mechanical Engineer with over 20 years of airport planning and major airport development experience.

Alexandre Bradley spent a number of years employed with the International Air Transport Association (IATA) as an airport planner manager. He was also the lead editor and 40% part author of the current 9th edition of the Airport Development Reference Manual (ADRM) published by IATA.

During the early 1990s the move to incorporate 100% Hold Baggage Screening at BAA airports was undertaken. Alexandre Bradley was the Project Design Engineer on the first ever 'in-line' hold baggage screening (HBS) 5 level installation at Glasgow Airport in Scotland. The systems and equipment configurations developed by Alexandre Bradley and Norman Shanks in Scotland have set the benchmark for the installation of HBS at all other airports across the world.

Alex Bradley has contributed significantly to the design of many airport projects around the world. Specific projects include:

- Stansted Major Departures and Arrivals Terminal and Satellites Expansion
- Stansted Generation 2 Project – second runway and terminal
- Prague International Airport – IATA Technical Advisor
- Panama City – IATA Technical Advisor Terminal and Baggage Systems
- Bangkok – IATA Technical Advisor
- Heathrow Terminal 3 HBS Installation (1995)
- Heathrow Terminal 3 Integrated Baggage solution (2009)

- Heathrow Shelterspan Upgrade
- Birmingham Eurohub HBS
- Edinburgh Phase 1 and Major Redevelopment
- Glasgow lines 1, 2 and 3 Expansion and Redevelopment
- Hong Kong Chek Lap Kok Apron and Baggage System
- Heathrow Terminal 4 Transfer Facility System
- Heathrow Terminal 4 Main Departures
- T1 Common User Lounge BHS System Integrator
- Gatwick Airport South Sorter Expansion Feasibility
- Gatwick North Terminal/Transfers and Mid Field Master Plan.

Alex Bradley has also been a BAA Technical Advisor to Reading University – Intelligent Airport Design Systems (1995–7) and a Technical Advisor to Transport Canada and CATSA (2002).

Acknowledgements

The author would like to thank the International Civil Aviation Organization (ICAO) for their support in the production of this book. The author would also like to thank the following individuals and organisations for providing case study examples:

- Questionnaire 1 – Tubari Alla, Head of Economic Analysis, Moldova Chisinau Airport
- Questionnaire 3 – Vince Scanlon, General Manager Airport Operations, Adelaide Airport Ltd.

Abbreviations and acronyms

ACI	Airports Council International
AMD	Archway Metal Detector
ATC	Air Traffic Control
ATM	Air Transfer Movement
ATO	Airport Ticket Office
BHR	Busy Hour Rate
BHS	Baggage Handling System
CAPEX	Capital Expenditure
CATSA	Canadian Air Transport Security Authority
CCTV	Closed Circuit Television (System)
CFD	Computational Fluid Dynamics
CT	Computer Tomography
CUSS	Common User Self Service
CUTE	Common User Terminal Equipment
dB	Decibel
DfT	Department for Transport
ECAC	European Civil Aviation Conference
EDS	Explosives Detection System – equipment designed to detect the chemical signature of explosive materials
EDDS	Explosive Device Detection System
EFTA	European Free Trade Association
FAA	Federal Aviation Administration
FAA Certified	Certification that indicates a manufacturer's equipment meets FAA performance standards
FAR	Federal Aviation Regulation

FIDS	Flight Information Display Systems
HBS	Hold Baggage Screening
IATA	International Air Transport Association
ICAO	International Civil Aviation Organization
IDL	International Departures Lounge
ILS	Instrument Landing System
IRR	Internal Rate of Return
LAN	Local Area Network
LCC	Low-Cost Carrier
MARS	Multi-Aircraft Ramping System
MCT	Minimum Connection Time
MPPA	Million Passengers Per Annum
OPEX	Operational Expenditure
pax	Passengers
PBB	Passenger Boarding Bridge
PLC	Private Limited Company
RAT	Rapid Access Taxiway
RET	Rapid Exit Taxiway
RZ	Restricted Zone
SSCI	Self-Service Check-In
TMP	Threat Mitigation Plan
TSA	Transportation Security Administration (USA)
TTS	Track Transit System
ULD	Universal Loading Device
URS	User Requirements Specification
Verti-sorter	Vertical Sortation Unit
WACC	Weighted Average Cost of Capital

Introduction to this manual

The objective of this manual is to provide airport planners with an impartial airport planning guide and reference material. The manual analyses the airport planning guidelines produced by the International Air Transport Association (IATA) and the Airports Council International (ACI) and makes independent recommendations; detailed analysis of the ICAO Annex standards is also provided. Associated airport planning software developed by the author is detailed under 'Related titles' on p. ii of this book. The manual comprises seven chapters.

Chapter 1, The brief to airport planners, defines the briefing material that airport planners need to take account of. The chapter explains the sensitivities surrounding forecasts, including upper and lower forecasts and the recommended medium forecast range to use. The chapter also explains the sensitivities surrounding the planning objectives for a small, medium or large airport. Service-level metrics are discussed and explained in the context of legacy carriers and low-cost carriers where operational expectations vary enormously. Development and phasing strategy objectives are also explained, as are the physical site constraints that can exist and how they can have a major bearing on the selected design strategy for an airport. As airport security becomes ever more important, the need for a competent airport design is paramount. The key airport security characteristics are explained together with the reasons why they must be addressed at the onset of design.

Chapter 1 also explains the forms airports can take, the operating functionality parameters and the analysis techniques that architects, engineers and planners will require to enable them to detail masterplan options that are innovative, effective, efficient and affordable to the airport client and airline users. To this end the characteristics of terminals, satellites and piers are defined. Most airports process cargo of one form or another. This chapter clarifies the location, size benchmarks and the general form of such cargo facilities. General aviation is often an important component of income for most airports. This chapter explains the briefing material that should be provided in this regard. Aircraft maintenance facilities and general airfield infrastructure are also explored and the reader is given the opportunity to use reference material for briefing purposes.

Chapter 1 also defines the briefing requirements to enable the airport planner to define the numerous support infrastructure and ancillary facilities that exist in an airport. Airport landscaping is critical to the successful design of most modern airports. The airport designer is directed to the clauses of ICAO Annex 14 which details the landscaping recommendations that can be used to provide an effective airport environmental performance. Finally, this chapter outlines some of the primary and secondary objectives associated with providing effective surface access provision, whether by rail, road or alternative modes of transport.

Chapter 2, Outline airport planning principles, examines and concludes an independent view of the IATA 10-step plan for masterplan development. A number of scenarios are explained, which focus on alternative types of runway alignment, primary and secondary airport development zones and general support infrastructure locations.

Chapter 2 also explains the various types of masterplan evaluation techniques that can be applied when assessing the validity and appropriateness of the master plans that may have been developed. Pairwise and weighting techniques are explained in full, allowing the designer to make effective design judgements based on a comprehensive evaluation technique. An example is given whereby the primary drivers for projects are outlined and then evaluated for a given airport development. Finally, this chapter includes a case study and a template that can be used for decision-making.

Chapter 3, Airport terminal and pier/satellite planning, focuses on the design of the terminal pier and satellite infrastructure. The complex problems associated with passenger segregation are explored and planning matrices are given to help explain the sensitivities that may exist at a particular airport. The benefits of process flow diagrams are highlighted and a case example is given to show how an airport operates with the complex, multiple, discrete passenger flows that exist. Levels of passenger experience and performance criteria are explained in the context of IATA service levels. The tools that an airport planning architect can use to determine whether the prospective design is effective are explored. These include passenger simulation tools and computational fluid dynamics software tools.

Chapter 3 identifies the key operational considerations that both airport owner and airlines need to consider. All airport developments have to be evaluated against key financial criteria. In this regard this chapter highlights the benchmark net present value and internal rate of return matrix that should be observed.

All airports will require efficient retail facilities to be provided to enable them to offer landing fees that airlines can afford. To this end airports need to provide effective retail facilities, whether they are landside retail or airside retail. The characteristics of these retail facilities are explained in detail. Case study examples are given of departures lounges in satellites and pier retail facilities, with explanations of income and rates of return for such facilities.

The following software is available to enable the core passenger processing and retail components of the internal space to be calculated statically: *Reference calculations for planning airport terminal buildings: a supplement to* ***The independent airport planning manual***. This software is available by contacting Woodhead Publishing Limited at: www.woodhead publishing.com.

Chapter 4, Airport baggage handling design, explains key characteristics of baggage handling design. The various categories of baggage handling system are explained, whether they are category A, category B or category C baggage handling systems as defined by IATA. This chapter tries to simplify the often complex characteristics witnessed within the baggage hall and explains the user requirements for baggage handling systems in simple terms. Manual handling requirements have changed enormously since 2004. This book explains the technologies that can be used and their pros and cons.

Chapter 5, Airport apron, runway and taxiway design, details all aspects of apron design with reference to the ICAO Annex 14 design standards. The function areas of the apron are explained with the design parameters that should be observed. The outputs from stand design tools are given. The optimum position and height of passenger airbridges are given for a wide variety of aircraft types. The advantages of configuring aircraft using Multiaircraft Ramping System (MARS) stand centrelines and the alternative types of aircraft parking aids are explored. At the end of Chapter 5 there is a summary of the characteristics of airports operating predominantly with low-cost carrier airlines when compared with airports that operate predominantly with legacy carriers.

Chapter 6, Design for airport security, focuses on the requirements to enable designers to develop safe airports, recognizing that existing airports can have poor security characteristics. This chapter looks at ways in which these existing airports can remedy these security deficiencies. The chapter shows how new airport developments can use modern information technology systems such as bomb blast simulations, vehicle approach speed simulation and rendered walk-through active simulations to create a safe design.

International, European and domestic security legislation obligations are explained and masterplan airport development considerations to meet those obligations are outlined. Chapter 6 also looks at the component parts of terminals, piers, satellites, car parks and forecourts and explains the best practice that should be adopted.

The differing security zones within the confines of the airport perimeter are defined. The various types of attack that airport or aircraft can sustain are examined and countermeasures explored. Alternative types of munitions and explosives are also examined and the impact that they can have on terminal infrastructure is discussed. Recommendations are given regarding appropriate countermeasures.

The blast zones and collateral damage that result from high explosives are discussed. The metrics that result from explosive simulation studies are given, as are recommendations on where best to site terminal infrastructure such as forecourts, building façades, car parks and any other zones considered to have a high population density.

Chapter 7, Case studies in airport planning, comprises a series of airport case studies. Two case studies look at the characteristics of airports processing predominantly low-cost carriers while three case studies look at characteristics of airports that process predominantly legacy carriers. This chapter has been developed with the kind cooperation of a number of airports across the world. Recognition of this cooperation is given in the acknowledgements section in this book. In each case study a template questionnaire was issued to the airport and the questions answered and documented verbatim. It should be noted that, since each airport is different, what may be best practice for one is not necessarily so for others. This series of case studies provides the designer with real examples of processes and layouts within airports.

1

The brief to airport planners

Abstract: This chapter reviews key overall issues in airport planning. It discusses planning objectives and target service levels. It then considers key requirements such as site constraints, construction logistics, airport security, terminal and satellite/pier design and cargo buildings. Finally, it reviews general aviation facilities such as those for aircraft maintenance, supporting infrastructure and ancillary facilities.

Key words: airport planning, security, terminals, cargo, maintenance.

1.1 Forecasts

All major airport developments require the provision of detailed and accurate forecasts. Significant airport development can take between five and ten years to develop from the onset to completion. There are various ways in which forecasts can be compiled, all of which will provide slightly different and slightly varying output data or forecasts. Forecasts are never absolute; they are calculated estimates of how the industry will perform in the future based on international knowledge, local knowledge and local business development planning for an airport. The industry is changing at a dramatic rate and some of the previously standard assumptions on traffic growth, market mix, etc., create great uncertainty and ultimately can lead to less than accurate predictions of how the business may develop at the airport. This creates a major dilemma for development managers who are trying to predict the performance of a speculative development. The business case evaluations for a multimillion pound/dollar/euro airport development need to have accurate predictions of how the income generated from incremental passenger growth is going to pay back the vast investment needed.

It is for this reason that no single forecast is relied upon and that a range of forecasts are considered. Typically there are short-term, mid-term and long-term forecasts for a major airport development. These can then be further divided into low throughput and high throughput forecasts. The forecast ranges are known to be estimates based on best industry knowledge and proven forecasting techniques; nonetheless, they are fundamental approximations of the likely traffic presented in a particular airport. It is good practice to assemble the forecasts into a booklet of data, which can be easily referenced and updated.

As a rule of thumb, assuming a common data intelligence source and metrics, it is reasonable to presume that the more immediate the forecast period the more robust the forecast is likely to be. This is purely because it is easier to understand foreseeable changes to the airline industry, which are affected by, say, political and economic stability, fuel prices and business and leisure travel industry trends. For this reason forecasts should be reviewed every six months and the metrics of a particular airport design challenged against the new forecasts range. A fundamental challenge may come from a pressure exerted by a local community group or by an air traffic control restriction. This could result in a challenge to the preferred mode of operation of the runway systems, where the segregated mode becomes a necessity over the more operationally flexible and traffic capacity enhancing mixed mode of operation. Examples of the forecast metrics that are typically included within a forecast dataset are explained later within this chapter.

When compiling the brief for an airport planning development those responsible for developing the forecasts should account for the following checklist:

- **Caution.** Forecasts are never absolute. They indicate what might occur based on gathered knowledge and intelligence of the airport operation and the industry going forward.
- **Forecast programme.** The programme of forecasts should be explained. This should detail what data will be available and when the regular updates are provided. This allows the key design decisions for the airport to be aligned with updated forecast statistics. This minimises abortive costs.
- **Market research.** The forecasting team should be actively involved in continuous market research. This will require gathering data from the airports in question in the local vicinity and industry knowledge gathered by attendance at industry symposiums such as, but not limited to, the International Air Transport Association (IATA) route scheduling venues. The sources of the data should be clarified along with all of the assumptions that go to make up the dataset. Governmental demographic trend models should be referenced, which typically detail population centres, profession type, age groups and forecasted disposable income streams for regions against future years. Further

sources of data would typically include Eurocontrol, International Civil Aviation Organisation (ICAO), Airports Council International (ACI) and IATA. All of these organisations provide commercial airline and airport industry specialised global views of trends in air traffic growth and change. Once this information is collated it can be used to provide local airport forecasts against assumption sets.

- **Forecast range.** The forecasters should consider providing specific data as defined within the following clause 1.1.1 of this publication. For the data type provided the forecasters should provide a lower, medium and upper forecast range. These ranges will allow designers to test the ability of their designs to cope with unforeseen lower and upper ranges in traffic.

Forecast material for a typical major airport development may require substantial forecasting data to enable it to be developed appropriately. Obviously the types of data provided for a small development, such as the development of a single satellite, will require far less data than the data required to specify the size and form needed to develop a major airport complex, comprising airfield runway, terminal buildings, fuel farms, car parking, and various types of modal traffic into and out of the airport. As forecasts go beyond five years it is also very difficult to predict the true nature of the traffic you are trying to forecast. It is for this reason that forecasts are often detailed forecasts for the first five years of an airport development. Beyond five years it is normal to provide only the key forecast metrics needed to detail the long-term aspirations of an airport. The payback period from major airport development means that forecasts are needed typically up to 25 years from the start of the project. The actual payback period can be as long as 50 years. Clearly it can be seen that a major airport development with the lifespan of 50 years could require a substantial set of data to support its analysis. Usually datasets every five years after the first ten years are used.

1.1.1 Forecast metrics

Forecast metrics critical for a major airport development would normally include the following parameters:

- busiest day period(s)
- air passenger departures, arrivals and transfer busy hour rates in the busiest day
- total annual passengers (MPPA (million passengers per annum)/split)
- passengers by market (international/domestic/business/legacy, etc.)
- passenger air transport movements by market
- aircraft movements by aircraft code

Table 1.1 Indicative forecast MPPA split between terminals (in millions of passengers)

Year	Total passengers	Terminal 1	Terminal 2
2020	15.0	15.0	0
2030	30.0	25.0	5
2040	40.0	30.0	10
2050	50.0	30.0	20

- hold baggage to passenger ratios by market
- hand baggage to passenger ratios by market
- aircraft stand demand busiest hour and busiest day by Aircraft Code
- cargo air transport movements and stand demand
- air cargo and mail (tonnage/type)
- on-airport employment
- on-airport employment reporting and finishing profile
- staff car parking peak space demand.

Smaller airport developments will require only component parts of this dataset to enable them to be of use to the designer in question.

Table 1.1 details the changing traffic split between terminals against design year. This high-level traffic split will confirm fundamental sizes of the infrastructure within the terminal complex. The period between 2020 and 2030 should be the initial focus for the forecasting team. It should be noted that Table 1.1 forecasts have been developed after the point at which the design team have concluded that a second terminal 2 is necessary rather than expanding the existing terminal 1. There could be many reasons for this, such as site constraints or operational restrictions. The forecasting team will not make this decision; this would be a conclusion reached by the design iteration to conclude the optimum design and capital expenditure (CAPEX) and operational expenditure (OPEX) cost model, which in turn will inform the forecasters.

Once it has been concluded that two terminals are required rather than one larger terminal, the transition period will need to be understood thoroughly. A key question will be 'Can airport infrastructure support the move of airlines that the forecasting team propose?' A demand versus capacity comparison will need to be carried out. A thorough understanding of the terminal occupancy, aircraft types and movements per hour demand versus the infrastructure capacity and operational capability will be necessary during this very sensitive period.

At the period when terminal 2 is proposed to be brought online it will be essential to understand the occupancy of both terminals and the downturn in traffic that could result in terminal 1. A number of terminal occupancy scenarios would need to be run to understand the sensitivities and the opportunities. Carriers do not like to have split operations as it is more expensive and disruptive for them, so a policy of keeping whole airline operations in one terminal should be received well by the airline concerned. It is also advantageous to keep airline alliances together; this promotes more efficient working practices for them and drives efficiency into the operation. This needs to be balanced against the cost to provide new infrastructure.

The bags to passenger ratios data source will be very important in defining the size and form of the baggage system. IATA provides a set of benchmark data (in the *Airport Development Reference Manual*, Table C2-1) and it is very wise to use local information gathered from discussion with the baggage system operatives and the airlines going forward.

As an example of the variations that can exist is the fact that international passengers travelling within the borders of Europe produce a ratio of between 1.0 and 1.5 bags/pax. This may be true of legacy carriers but low-cost carriers are actively aiming to reduce these bag to passenger ratios as much as possible. This allows their aircraft to fly light, to provide timely departure and to ensure that landing charges are kept to a minimum by reducing the need to support spending capital within the airports on complex baggage system projects. Low-cost carriers for the same routes would look to have a ratio of between 0.5 and 0.75 bags/pax. This single variation can have a massive impact on the size of the terminal building and complexity of the baggage system and the resultant capital cost of the infrastructure.

It is the recommendation of the author that historic bag to passenger ratios in this regard are understood. Future bag to passenger ratio aspirations should be recorded through detailed discussions with the airlines resident at the airport. This information used in conjunction with forecast information and business development information should help to estimate the necessary baggage system and terminal size that should be safeguarded. The key is to provide flexible infrastructure that you really need when you really need it and not before. It is important to have the flexibility to adapt the infrastructure and systems should the traffic forecasts vary enormously.

Figure 1.1 shows the daily passenger arrival profile and the forecast mid-range passenger departure schedule for airports processing 35 MPPA. Both graphs show that the early morning peak exists at 0800 hours and that there is an afternoon peak at 1830 hours. It is important to note that the airport with this profile has the capability to process 35 MPPA and the resultant busy hour rates of the 95th percentile equates to 3800 passengers per hour.

Conclusion: Fig. 1.1 graph
35 MPPA – promotes, say, design option A

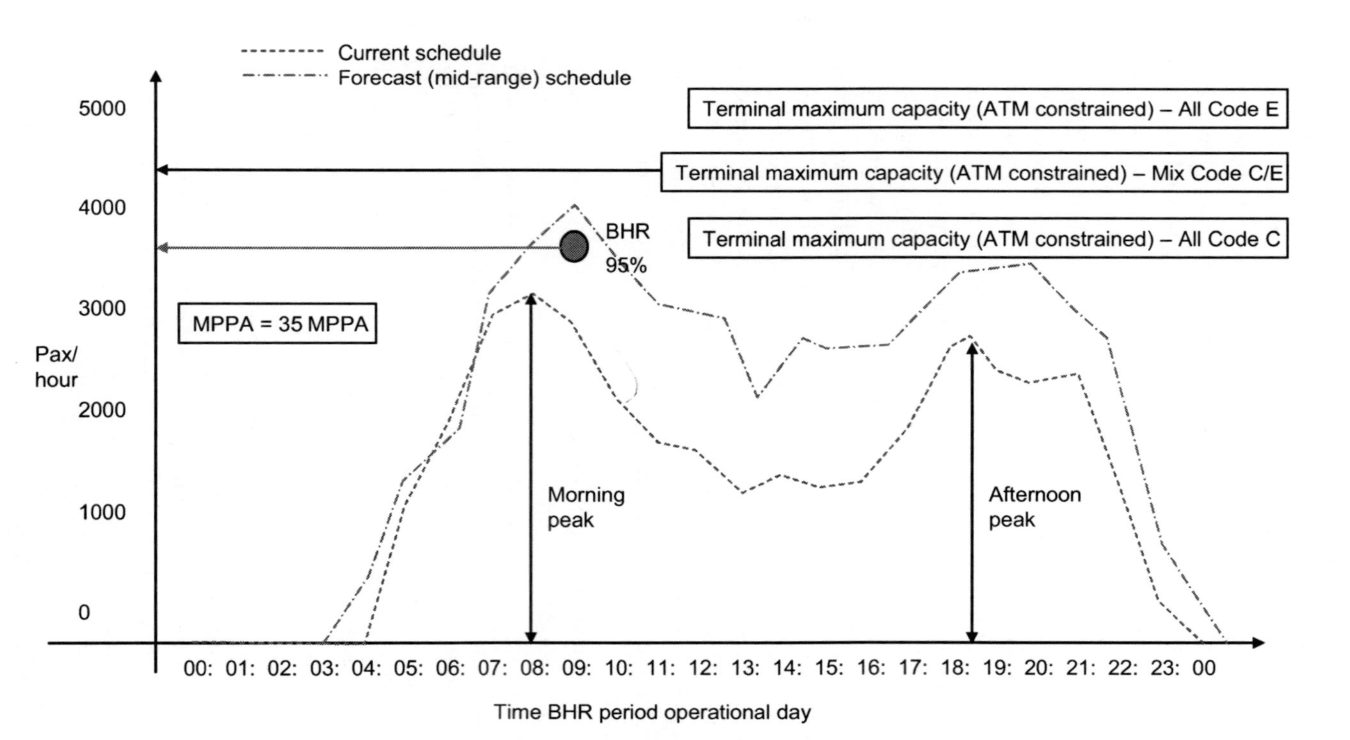

1.1 Forecast reference data (for example only; subject to airport specifics).

Table 1.2 Results of 35 MPPA by different busy hour rate and resulting areas required

Option A	35 MPPA	BHR 3800	Resultant area ~47 800 m^2
Option B	35 MPPA	BHR 4750	Resultant area ~49 700 m^2

Peak 4000 pax/h
95% busy hour rate (BHR) = 3800 pax/h
Busy hour grow (AM and PM)
BHR growth (AM) assumes Code E component
Shoulder period grow – more terminal redundancy needed

Figure 1.2 shows the daily passenger arrival profile and the forecast mid-range passenger departure schedule for airports processing 35 MPPA. The forecast graph shows no early morning peak or afternoon peak. Instead the forecast shows the lunchtime peak significantly higher than denoted in Fig. 1.1. The airport with this profile also has the capability to process 35 MPPA and the resultant busy hour rates of the 95th percentile equates to 4750 passengers per hour.

Conclusion: Fig. 1.2 graph
35 MPPA through still – totally different design B
Peak 5000 pax/h
95% BHR = 4750 pax/h
Busy hour grow (AM and PM) – lunchtime wave more significant
BHR growth (AM) assumes more Code E component
More Code E = bigger pulses of passengers going through infrastructure
Shoulder period significantly grows – more terminal redundancy needed

Consider two different scenarios: both airport designs are for processing an annual throughput of 35 MPPA, but the busy hour rates change from 3800 to 4750 pax/h. What is the likely change in the area needed to process these passengers? Table 1.2 summarises the results in the area required to process different busy hour rates. It should be noted that there are numerous metrics required to define the size of the terminal facility. These can include quality of space check-in options, queue depth requirements and retail variables, to name but a few.

Important note. The areas denoted in Table 1.2 have been calculated using a terminal planning model which has numerous other inputs over and above the busy hour rate which help define the requirements such as, but not limited to, traffic type, staff and equipment processing rates, etc., to name but a fraction of the variables to be considered. It is important to understand that the areas depicted in Table 1.2 have used constants for all

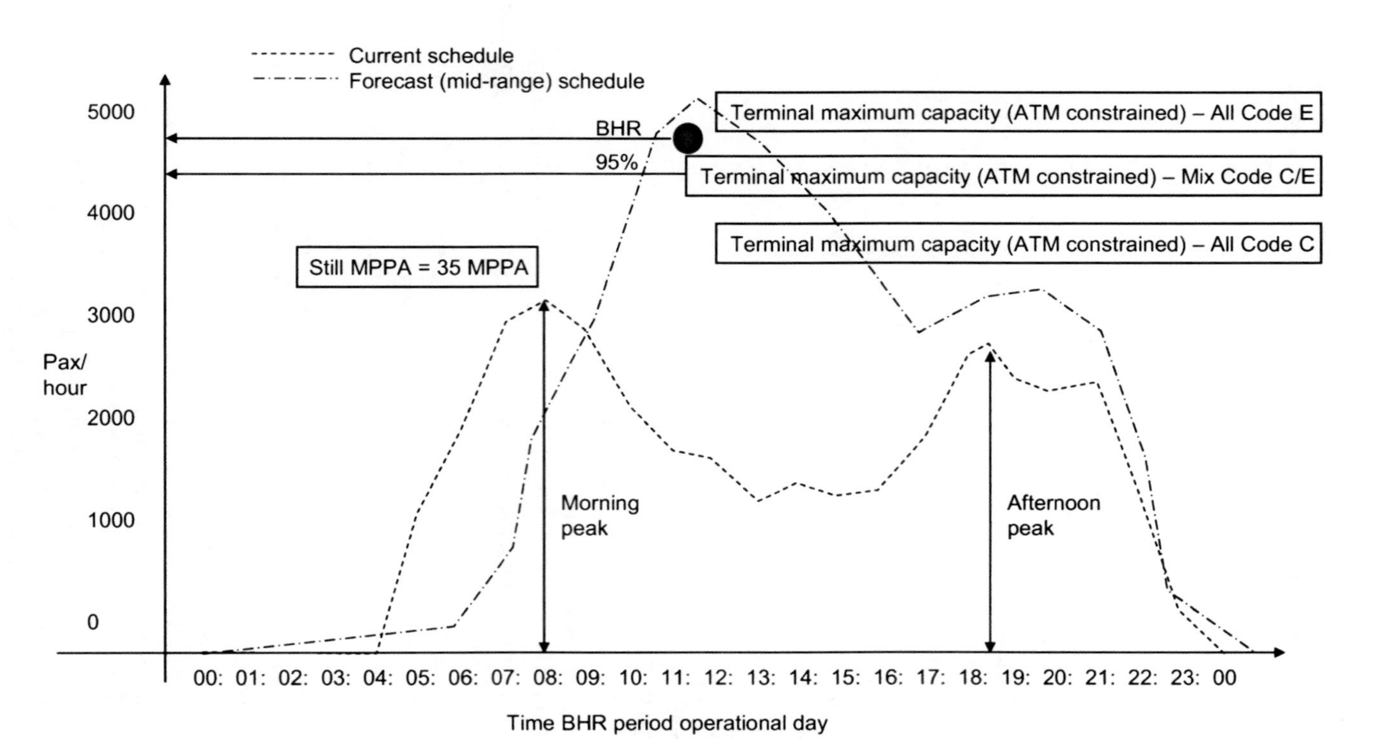

1.2 Alternative forecast reference data (for example only; subject to airport specifics).

these other input parameters and that only the busy hour rate has been varied.

In this one simple example, the impact of one change in characteristics of the forecast, such as the all-important busy hour rates, can dramatically change the size of this facility and its resultant cost as well as its ability to adapt to change.

1.2 Planning objectives and service levels

At the onset of a project it is important to note the planning objectives and service level objectives that the development should seek to meet. Whether it is an existing airport or a new airport development, the planning objectives and service level objectives will need to be documented.

There are very few completely new airport developments. The planning objectives for a major new airport will need to sit correctly within the strategic national airport development plans. These strategic airport plans will take into consideration regional expansion for housing, rail connectivity, road collectivity and employment opportunities in the longer term, namely a 25-year plus horizon. Additionally, strategic national airport infrastructure development plans will need to take full account of the air traffic route capability available in the short term and those that can be developed in the future. Considerable attention will be needed to understand the full impact of those air traffic routes on both the commercial viability of the routes with respect to airlines and passengers and the environmental concerns that can be created.

In the context of a major airport development and on the working assumption that the strategic national airport development plan supports the development of the airport, the airport development team can begin to develop multiple options to evaluate the best airport solution. At the highest level the runway alignment and length of the runway will need to support the strategic national airport development plan. There will be options and variations associated with runway length and to some extent with runway alignment also, as fundamentally they should be aligned to the strategic national airport development plan. Runway length will be a function of the makes of aircraft forecasted to use the airport, giving due consideration to the strategic views from neighbouring villages, noise considerations, proximity of the runway and taxiway relative to the terminal infrastructure, the elevation of the runway and land mass introduction or removal, and finally and very importantly general environmental concerns such as ecological considerations.

Understanding the impact of noise pollution is critical to the success of any airport planning project, particularly those projects that involve the development of runways, taxiways or apron areas where aircraft noise will likely change the general noise signature of the airport. Once a series of masterplans have been developed, which meet the business needs, it will be essential to understand the noise contour characteristics of each of the masterplan developments proposed. A specialist in noise and noise abatement will need to be employed in order to understand fully the noise characteristics of the solutions and how best, through the use of effective design, to mitigate the impact to within a tolerable dB level and a geographical and economically viable set of metrics. It is common practice to articulate a set of noise contours for the total airport development where the 63 dB leq is the most critical noise level to be considered as tolerable. Noise levels above the 63 dB leq are generally considered to be the position where compensatory or land purchase schemes are initiated, though variations to this rule of thumb will occur on a discretionary basis.

Air pollution shall need to be considered thoroughly where the influences of aircraft and terminal infrastructure CO_2 within any proposed design solution will need to be understood and documented. Computer fluid dynamics (CFD) software coupled with competent users is recommended as this can simulate the effects of air pollution propagation. There are four main epicentres of aircraft CO_2 pollution within any airport complex, which include:

- runway V1 full throttle locations
- concentrated aircraft stand positions (apron areas)
- taxiways
- departures and to a lesser extent arrivals flight paths
- engine test bed areas.

The airport planner must consider the planning sensitivities associated with the security performance of any chosen masterplan. The reader should refer to Chapter 6, Design for airport security, which explains what should be examined in this regard. A thorough risk assessment of the airport security characteristics needs to be understood and taken into account.

It is necessary that the views of the thresholds of the runways can be seen unaided from the control tower observation control rooms in normal good weather conditions. It should be noted, however, that the location of the control tower should not take precedence over the far more expensive location of the runway.

Runway location will be heavily dependent upon aircraft noise characteristics. Noise dB concentrations can be difficult to predict and it will be necessary to employ computer modelling to understand the likely effects of the chosen wider airport planning solution using a noise modelling tool and a competent specialist. Mitigation measures required to satisfy the environmental noise concerns will be necessary, such as, but not limited to, land mass modifications as defined in ICAO annex 14. These measures will be necessary but can be very expensive to implement. It is for this reason that multiple options should be considered for the runway in terms of taxiway placement and holding areas and precise placement of the runway. It is unlikely that when all of the objectives of the performance, capital cost, operational safety and operational efficiency have been taken into account a single solution will be identified. It is far more likely that many of the options that are created will have varying degrees of acceptable performance and will require an analysis comparison so that all of the needs of the development have been appropriately and objectively analysed. It is very important to gather intelligence of the concerns of the local inhabitants to a radius of at least 8 miles. A good mechanism for gathering this intelligence is to conduct public consultation forums. These can take the form of ongoing development proposals produced by the design team, which can then be discussed with the local residents. The resultant documented concerns can then be used to help form the assessment process and priorities associated with the performance of characteristics such as noise flight path, take-off and landing patterns, issues concerning blight and compensation claims. The output from these public consultations can very effectively provide guidance to the design team to enable them to focus their designs on proposals that carry the least impact and are therefore more likely to proceed through planning permission along the path of least resistance.

When considering the planning objectives for a major airport terminal alone on an existing airport complex there will be natural restrictions to the location, height, form and functionality of the building relative to existing infrastructure. In the same way the location of piers and satellites on the existing airport will heavily influence the location of the new pier and satellite infrastructure. In the context of the terminal building the key factors, deliberately not in order of preference, that will need to be taken into account when developing the new terminal will include:

- the primary functions of the building – departures, arrivals and transfers
- local planning constraints set by the planning authority
- elevation of the building and impact on strategic views from local inhabitants
- structural and glazing restrictions
- soil mechanics
- collectivity with adjacent piers, satellites and connecting modes of transport systems
- capital cost
- operational flexibility
- operational cost
- environmental performance

- operational efficiency: airline and airport operations including ground operations
- operational performance with respect to connecting times of particular importance for transfer hubs
- ability to provide adequate surface access capability – rail, road, ferry, etc.
- suitability to provide cost-effective car parking solutions (short-, mid- and long-term products)
- provision of suitable space to accommodate airport operator staff, airline/ramp staff, maintenance staff, control authorities staff, police and general landside facilities such as, but not limited to, aircraft catering terminal retail deliveries.

The above items list some the external factors associated with the location of the terminal building. The precise internal configuration layouts and levels (single or multiple) shall be dependent upon very precise characteristics of the medium- and long-term operational requirements stipulated by the forecasted airline and operational needs. The medium- to long-term aspirations for the airport dictated in the forecast shall heavily influence the internal configuration of the building, the size of the building in its initial form and how it is migrated into a longer-term solution.

The following key questions should be asked:

- Question 1. Is the airport masterplan capacity objective set to the national strategic airport development plan?
- Question 2. What is the airport traffic type to be accommodated – departures, arrivals, transfers, domestics, low-cost carrier, domestic traditional, home-based low-cost, away-based low-cost, scheduled domestic, scheduled international, charter short haul, charter long haul, etc.?
- Question 3. What availability of capital is there?
- Question 4. Are there any airport regulatory restrictions and long-term strategic safeguarding requirements for regional air traffic growth that need to be accounted for?
- Question 5. What are the environmental targets for the airport? Are they airport owner targets or national governmental targets?
- Question 6. What are the baggage processing requirements?
- Question 7. What are the connecting times requirements (departing only–transfer hub), passenger to gate?

Once these fundamental questions have been answered, this will permit an effective project brief to be written.

1.3 Development phasing strategy

The development phasing strategy is a description or story board of how the airport complex should evolve and progress. The phasing strategy should focus on and be capable of addressing the major likely scenarios that could occur and should confront the airport operation. The phasing scenarios shall need to understand:

- **Forecast demand.** What happens if forecasts show large growth or material changes in traffic type, long haul to short haul, point to point becomes transfers, etc. The development phasing strategy should show how benchmarked developments capable of dealing with these scenarios need to be accommodated physically on a layout. The precise details within these facilities do not need to be described but they do need to be capable of being built.
- **Buildability.** The location of structures and sequence of developments need to be capable of being built using established and proven construction techniques. It is of no use to put forward a solution that cannot be constructed. The location of the structures, and to a lesser extent the apron infrastructure, need to take into account surface access of materials to the sites and waste materials from the sites. All of this needs to be achieved while keeping the airport fully operational. This requires meticulous planning for each phase of development.
- **Operational issues.** A major issue for airports is the need to provide terminals, piers, satellites and apron infrasructure with the correct client/passenger/aircraft occupancy. As an example consider the development of a major airport terminal building development alongside existing terminal buildings. The existing terminal will have an established client base, which may need to be migrated in part to the new terminal building to create an effective operation on opening. The development phasing plan will need to define what traffic types move, and when and what impact it will have upon the commercial and operational viability of the airlines in all terminal facilities. As a principle, airlines do not like to have a split check-in operation, so migration plans need to have short-term and long-term flexibility.
- **Efficient capacity.** A major requirement is trying to understand what is the appropriate amount of capacity provision that should be provided in each phase. As a principle a 'just in time' philosophy is the best to use; although there are some good rules of thumb to adopt, all have pros and cons. For established airports, which have a steady year on year requirement over the next 25-year forecast period to expand by XMPPA, any single terminal/pier development should try to limit the step increase in capacity by XMPPA/5 build increments. As the factor 5

is reduced, so the commercial risk is increased as more capacity is built into any single step. All scenarios should be backed up with an evaluation of financial returns for phases. The exception to this is the commercial development of new runways, although runway length expansion should be equally assessed for commercial and operational need with a 'just in time' capacity build philosophy.

1.4 Site constraints: physical and operational

The planning team will need to understand the physical and operational site constraints that will exist for an airport development location. The physical site constraints will extend to a boundary appropriately sized to facilitate the forecast and should not be limited to the constraint of the current ownership boundary. The operational constraints will extend far beyond the physical site constraints and shall be a product of the air traffic and noise abatement solutions proposed for any development. As an example, a series of flight path proposals coupled with noise abatement plans, defined outside the physical airport boundary, will dictate how runway, apron and terminal infrastructure will be permitted to be used operationally within the airport site boundary.

If the airport is to process more than 100 MPPA in 25 years and currently occupies land where only one runway resides, then the potential ownership boundary will need to be extended to line up with the likely building and expansion zones. The definition of the new boundary will be the product of detailed operational and financial scrutiny and will need to take into consideration such items as (though not limited to): dB noise levels, construction issues, land purchase costs, compensation boundaries as well as protected land masses, sites of environmental sensitivity, historic sites and issues concerning strategic views of airport infrastructure from key villages, towns and cities. The site constraints will need to take into account mandatory ICAO runway and taxiway separation distances and length of runways, all of which can have a major impact on the land take required (refer to ICAO Annex 14, particularly Chapter 3). There will be third dimensional restrictions such as a restricted surfaces requirement, e.g. power lines. There will be airport buildings and infrastructure that it would be difficult to justify changing, either financially and/or operationally; these will need to be defined as constraints and categorised accordingly. Constraint categories can be defined as in Table 1.3.

1.5 Construction logistics

There are many construction logistic issues to consider when embarking on the development of an element of airport infrastructure. The airport

Table 1.3 Physical site constraint categories

Bradley constraint category	Constraint type	Expiry date of sensitivity
High (RED)	Financially sensitive Asset not life expired (high residual asset value remaining) Operationally sensitive Third party land ownership – not willing to sell	> + 10 years
Medium (AMBER)	Financially sensitive Asset not life expired (high residual asset value remaining) Operationally sensitive Third party land ownership – not willing to sell	> 5 years ≤ 10 years
Low (GREEN)	Financially sensitive Asset not life expired (high residual asset value remaining) Operationally sensitive Third party land ownership – not willing to sell	≥ 1 year ≤ 5 years

complex should be broken down into those developments that are airside and those that are landside. The demarcation of projects can be broken down further to suit the size of projects that are likely to occur so that work streams are equally distributed.

Projects that reside well within the airside environment will require careful planning so that materials and staff can access and egress the site with due consideration to the site security requirements. Staff security access permission in these areas can be a very problematic issue, as it can be difficult to obtain the necessary security passes for a less technically well-qualified workforce, which often change employers frequently. If at all possible it is preferred to make these airside building sites temporarily landside during construction, wherever practically possible. This can be costly in its own right, however, as the restricted access boundary will often need stronger walls, temporary closed-circuit television (CCTV) and other measures to ensure that the relocated boundary line is secure.

The positive side of working within the airside environment is that materials and tools used in the construction process are far more secure and less prone to be stolen. Site security costs are also significantly reduced.

Larger more major construction projects such as the building of new runways, taxiways or terminal buildings need to address far bigger, more environmentally sensitive issues. These can include:

- **Earthwork balance.** Confirming what the earthwork balance strategy will be. When building an airport terminal building or runway the geography and topology of the ground varies enormously and ground levels can oscillate tremendously. Moving the runway 5 m to the left might not have a major cost impact in terms of operational and land purchase issues, but could create a major cost and environmental impact concern. All the options are therefore developed so that they can be examined and evaluated and the best overall option can be chosen.

- **Surface access plan.** The volume of materials that will need to be accessed and be removed from the construction site could be enormous. A complete and well-understood surface access plan will be required, which should identify traffic type, traffic routes, traffic frequency, safety and the noise and pollution impact during the construction period. All of these factors need to be developed with the local planning authority and local resident representative forum buy-in wherever possible.
- **Operational restrictions.** When building any type of airport development there are likely to be operational restrictions associated with the work that is being carried out. In the context of runways and taxiway and apron/stand developments, the restrictions can be extensive. The sites need to be hoarded off and the material on the apron secured so that materials are not accidentally digested by aircraft engines, and no damage is caused to aircraft when debris is flicked up from aircraft landing gear while moving, or by support vehicles. The operational teams will define the periods when construction can take place. This will ensure that ground staff and airlines are made fully aware of the necessary working practices that are needed.

Figure 1.3 shows the construction sequence used for a major terminal building extension. The sequence of build and placement of the crane is shown along with the operational restriction imposed upon the construction team as a result of the location of passengers and the location of the crane hoist over passenger zones.

1.6 Airport security

The designers need to understand the security legislative parameters that they should be working within. These are also likely to change throughout the course of the project. At the highest international level the security requirements are set by those states that subscribe to the aviation Annexes of the ICAO. The security parameters are contained within ICAO Annex 17. These ICAO requirements are then often reinterpreted by the national government in question, giving a bias towards the national/domestic situation and perceived threats and risks to aviation in that sector. In the context of airports and airlines operating from within the European Union these groups subscribe to both ICAO and European Civil Aviation

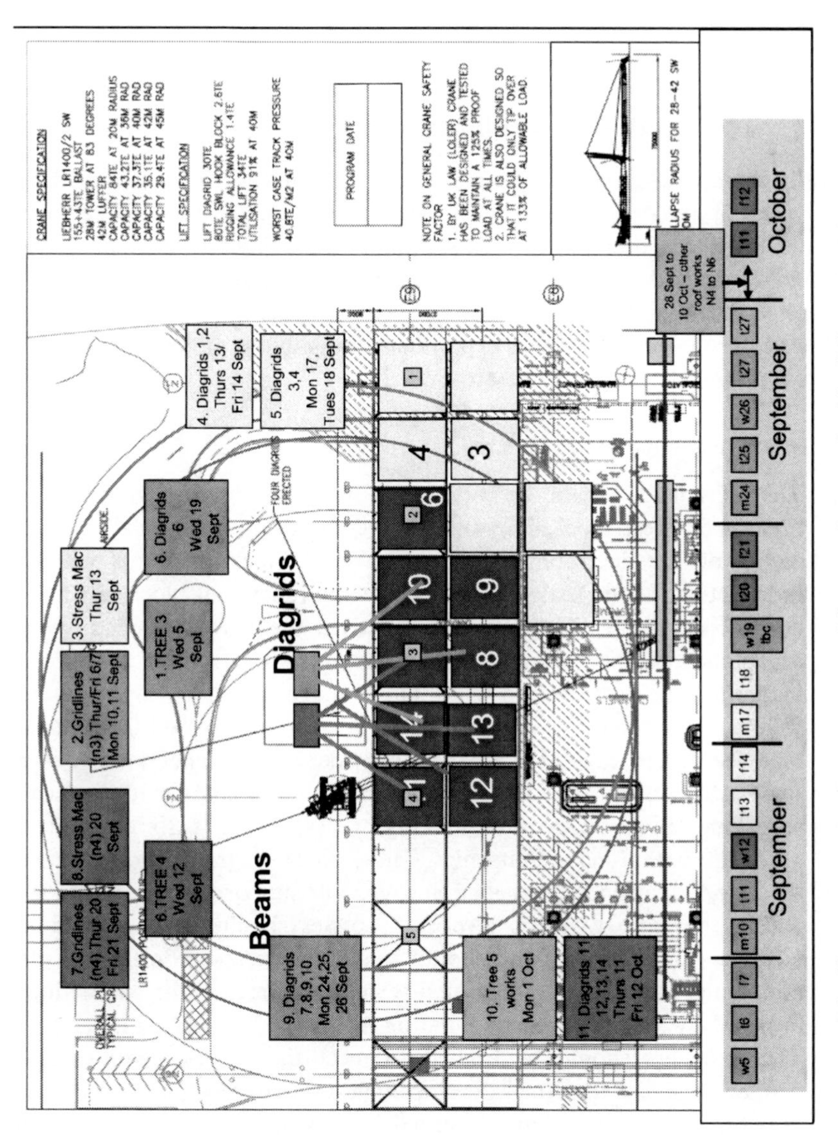

1.3 The construction sequence used for a major terminal building extension.

Conference (ECAC) documentation documents 30 and 2320, now replaced by EU300. ECAC is the European subdivision of ICAO. In the United States of America they refer to the Federal Aviation Authority (FAA) and for security the Transportation Security Administration (TSA) mandates, processes and documentation. The 1963 Tokyo, 1970 Hague and 1971 Montreal ICAO security conventions are also extremely important to note.

1.7 Terminal and satellites/piers

Ideally the architect, engineers and planners should not be constrained by very prescriptive definitions of what the client terminal and satellite form should be when compiling the masterplan brief. These groups should be steered towards developing the building that best meets the operational, financial, flexibility and environmental needs of the client. The client should allow the professional teams to explore new boundaries of design that meet these fundamental needs of the airports.

Important requirements:

1. The designers should be briefed initially to develop block level diagrams. These block diagrams should depict the functional requirements of the development. The first level block diagrams will be independent of the scale of the need of the functional block but must define basic connectivity explanations. Once these are developed to the acceptance of the client it is then possible to introduce a scale of magnitude into the proposal of block diagrams. It is important to follow this sequence so that the basics are well understood first and so that the design can be audited to provide correct functionality. The scale attribute if added too early can introduce a complexity that can sometimes cloud the decision making process; hence it is important to agree the basic functionality first. These block diagrams should define the airport passenger, baggage and general support processes needed within the airport. These airport processes should be developed in advance of any layouts and should enable the designers to understand and agree the desired functionality without having the restrictions imposed by architecture and building location.
2. The designers should be briefed to meet the process requirements defined in step 1 and then develop high-level concept plans of all the possible locations of the terminals, piers and satellites that will technically meet the forecast, service level performance criteria, international and national design standards and international recommended practices.
3. The designers should have the maximum and minimum building height

constraints defined. Each of the core buildings should have its primary function (international departures/arrivals/transfer, etc.) and its secondary functions defined. Segregation issues should also be identified (e.g. Schengen/Special).

Definition. The restricted zone (RZ) boundary is the secure demarcation boundary which separates landside unscreened passengers, staff, goods, baggage and process zones from the airside secure and screened zones of the same type.

It is essential that the process block position of the RZ for a development is defined at the earliest opportunity, e.g. a first pass (step 1 above) block level requirement should contain the RZ block level position as this will influence the process blocks either side of the RZ.

1.8 Cargo buildings

Most modern airports have the capability and the need to process cargo whether it be belly hold cargo or full cargo including express cargo operations. The brief will need to set the strategy for dealing with these cargo models. It is often useful to look at the location issue from two perspectives before deciding where the best fit is. To this end the brief should stipulate the traffic flows to accommodate the cargo. There are likely to be a limited number of preferred road and rail routes into and out of the airport for cargo process. These preferred routes should be highlighted clearly to the designers and where they intersect with the RZ boundary should be clearly identified (potential cargo road/rail routes). Then the cargo air traffic should be identified in the forecast. The ICAO coding of aircraft should be understood and the daily demand mapped on to current stands for an existing airport or preferences given to locations where the cargo stands could be best located. In the context of cargo aircraft operations these are often Code D and above sized aircraft. If, for example, the wind direction is predominantly from one direction, then the cargo stands are arguably best located closer towards the rapid access taxiway for the predominant take-off direction. This is because far less fuel is spent moving a fully fuelled heavy aircraft to get to the runway holding point. It should be noted, however, that this simplistic operationally biased view is not necessarily the final best location for the placement of the cargo stands. There are a number of other factors that need to be taken into consideration, such as: land availability, environmental impact and noise. Often cargo operations run throughout the night and these operations carry significant noise issues. In the dead of night the loading and unloading of bulk goods can be very intrusive from a noise perspective. When the preferred cargo stand and cargo processing building location are understood from landside access and from an airside perspective the best location solution will probably be where the preferred road and rail access routes converge with the preferred stand locations. It is often the case that a compromise is taken when all these factors have been considered.

At masterplanning level the size of the cargo buildings will be less well defined but the cargo volumes should be forecast. The following benchmarks are a good guide to enable the size of the cargo terminal buildings to be estimated:

- hub: 5–7 tonnes m^2 per annum
- predominantly manual operations: 5 tonnes/m^2
- some automation: 10 tonnes/m^2
- fully automated: 17 tonnes/m^2.

The setting out areas should be briefed, giving reference to:

- depth of cargo building 60–90 m at least
- cargo terminal edge to unit loading edge: width 6 m
- cargo road: at least 12 m width
- staging area: cargo road edge to nose of aircraft 18 m width.

1.9 General aviation facilities

Often airports that process less than 2 MPPA will have a sizeable component of general aviation. These are defined as those aircraft movements confined to light leisure aircraft and executive jet/prop type operations. Where possible it is best to process all passengers, crew and pilots through the main terminal building for financial reasons. There are many occasions and instances where the general aviation facilities are separate buildings. In this regard it is best to centralise all security screening processes, as equipment and staffing is expensive.

Where it is deemed more appropriate to separate the main passenger terminal operations from the general aviation facilities then it will be necessary to create a general aviation facility that is appropriately sized, containing an almost mini scaled-down replica of the main passenger terminal with the minimal passenger and baggage processes. The location of the facilities will be influenced by the income achieved from the facility and by the stand capacity. The terminal will need to inhibit the same security characteristics as the larger main terminal building with full restricted zone boundary integrity and operational security processes. If the general aviation traffic is scheduled to grow it will be important to align this to the main passenger terminal traffic growth plans.

It is common to separate the general aviation facilities for security reasons as it is often perceived that general aviation is subjected to lesser screening technology and as such general aviation aircraft and passengers and flight crew are kept a sizeable distance apart. It will be a balance between optimising operational costs to a minimum versus available land versus optimising security characteristics.

1.10 Aircraft maintenance facilities

All airports need to permit some form of aircraft maintenance. This can be as simplistic as a remote stand area with limited store holdings and tools to the fully integrated maintenance hangars provided at large gateway airports. It will be important to speak to all the resident airlines and to those forecast to be resident and then make plans accordingly to provide the type and capacity needed. A major factor that influences whether or not major maintenance is carried out at an airport will be the local cost of highly qualified staff.

Ideally aircraft maintenance facilities should be located away from the terminal building as the work undertaken often carries risks that, if located close to the terminal, would present unacceptable dangers. Also the maintenance work can be noisy so it will be helpful to locate these facilities in zones that are away from the terminal and in places that do not create problems to local residential areas. A particular concern is the location of engine test beds, which can be very noisy and quite dangerous sites. Larger airports may require multiple maintenance facilities so as to provide extra redundancy resilience.

The size and complexity of the maintenance facility will be dependent on many issues:

- demand/need from the client base
- availability of land and noise mitigation measures
- suitably low operating cost
- quality and availability of suitably qualified staff
- reliability and availability of spare.

The following parameters should be defined in the brief to the designers:

- Confirm the code of aircraft that need to be maintained.
- Confirm the current and future client base.
- Define the hangar space required as a ratio of forecast air transfer movement (ATM) unless specific (m^2/ATM).
- Define the preferred location of hangars.
- Confirm any existing hangars and operational limitations of their use.
- Define any specialist noise mitigation and drainage requirements associated with aircraft engine test rigs and general aircraft maintenance requirements.

1.11 Airfield infrastructure

When defining the briefing requirements for the airfield infrastructure the following parameters should be defined and in each case the m^2/ATM should be confirmed:

- airport operational management suite (airline scheduling briefing management and rest accommodation)
- fuel supply/depot/primary and auxiliary power supply for terminal and ramp power
- ramp staff accommodation
- ramp vehicle parking
- de-ice facilities (if drive through is used)
- fire station space and processing needs (ICAO Code definition explained); special care is needed to comply with various categories of fire response coding and facility size and placement (refer to ICAO Annex 14 Clause 9.2).
- specialist areas:
  - technology requirements used for intruder detection across perimeter and within restricted zones
  - anti-terrorism aircraft stands.

1.12 General support infrastructure

When defining the briefing requirements for the general support infrastructure, the following parameters should be defined and in each case the m^2/ATM should be confirmed:

- control tower location and operational restrictions
- baggage hall sortation system
- baggage tug charging facilities
- baggage universal loading device (ULD) container storage facilities
- apron equipment – towbars, stairs, etc.
- mobile power units
- mobile fuel delivery truck
- control authority vehicles
- passenger trolley distribution processing areas
- retail storage and security screening facilities (internal within the terminal or off site)
- national mail storage areas
- car hire facilities and vehicle parking
- car parking (fast track/short-term/medium-term/long-term).

1.13 Ancillary facilities

When defining the briefing requirements for the ancillary airport facility infrastructure the following parameters should be defined and in each case the m^2/ATM should be confirmed unless stated below:

- terminal and apron water treatment processing units
- storm water balancing ponds for car parks, runway and apron
- aircraft catering units
- police services:

Pax PA/MPPA	Full-/part-time	Manpower needs
1	Part-time	2
5	Part-time	4
10	Full-time	7
30	Full-time	15
70	Full-time	35

- national immigration services
- taxi rank pre-make up
- airport maintenance teams
- fire training ground.

In accordance with the intent of ICAO Annex 14 clause 9.2, the primary role of the airport fire services is to:

- Provide a 24 h 365 days a year fire extinguishing service to the apron and terminal areas only.
- Monitor the apron and terminal areas for early signs of fire and prevent accordingly.
- Assist ambulance and police departments in the event of airside and landside vehicle crashes.

1.14 Landscaping

The brief to the designers should contain a full description of landscaping objectives, which give reference to the following items and include:

- Confirm the percentage of green belt land to be provided in a phased solution.
- Define key strategic views.
- Confirm that the masterplan uses engineered land bank bungs to mitigate noise or line of sight issues or mechanical structures.
- Noise contour objectives – identify areas of sensitivity – confirm where the sensitive zones are.

- Define the use of plants for adverse noise and line of sight mitigation – ICAO recommendation to mitigate bird/wildlife attractants.
- Excavation versus land balancing objectives. What is the target volume of land to be removed versus land to be bought to site.

1.15 Surface access

Surface access provision will be key to the success of an airport. If it is easy to gain access to the airport then it is likely to be very favourable to the long-term success of the airport. Obviously to get to this 'Nirvana' position the designers will need to evaluate all of the access option permutations and cost these against the evaluation criteria discussed in Chapter 2 of this book.

When defining the briefing requirements for the surface access provisions the following parameters should be defined:

- forecourt access
  - passengers' walkway access, e.g. terminal to terminal routes not to exceed say 200 m and not to undergo two level changes
  - confirm the number of vehicle lanes required and their use (bus/car/taxi) and dwell/parking zones/distance from terminal infrastructure and blast protection measures
- passenger car and airport car park bus routes – routes and number of lanes required and parking and dwell periods
- train/tram routes
- cargo vehicle routes
- ferry boat access
- cycling access and foot passenger access
- rental car routes (call on demand or local? – former – think safety)
- goods vehicle and waste disposal vehicle routes
- confirm limit of breakages in existing roads network.

Once you have developed an airport brief that follows the guidance detailed in this chapter a comprehensive, flexible and not overly restrictive brief will have been compiled. The brief should then be managed and controlled. Forecast changes and legislative changes will inevitably come along and the project brief shall need to respond to this and be issued accordingly. The project brief should then be issued to the design team who should refer to Chapters 2 to 6 inclusive of this manual to ensure that the airport development progresses appropriately.

2
Outline airport planning principles

Abstract: This chapter reviews key airport planning techniques and drivers. It discusses the IATA planning best practice 10-step plan as well as the use of evaluation criteria, pairwise and weighting techniques. It then reviews performance drivers such as passenger experience and flexibility of operations as well as operational, financial and environmental performance. Finally, the chapter includes a case study to show planning principles in practice.

Key words: airport planning, IATA, passenger experience, operational performance.

2.1 Industry standard planning approach

The IATA *Airport Development Reference Manual* defines the generic masterplan sequence that should be adopted for essentially green-field site developments. This masterplan sequence fits both passenger airport developments and cargo airport developments by being able to exchange steps 7 and 10 accordingly. This chapter explains the IATA 10 steps and provides concise supplementary information. When defining the masterplan for an existing airport development, which may have already been in operation for 20 or more years, the sequence is likely to need to change, owing to existing knowledge and/or pressures unique to the airport location. For example, a passenger airport that has considerable runway capacity and fixed runway constraints would be likely to have the runway alignment defined in IATA masterplan step 3 as a 'fixed constraint', which cannot be changed. In contrast a green-field site may require considerable evaluation of all the runway alignment options at an early stage to determine the best location of the runway(s).

It is important to understand the constraints that exist and to really challenge if they are constraints. While in the previous paragraph it was noted that an existing runway was likely to be a constraint, the value of keeping it a constraint should be explored. The challenge is then to understand the operational impact and resulting operating cost and revenue generating loss of such a constraint when compared with a new runway position. The knowledge that should be addressed for existing runways should include:

- What is the condition of the existing runway (as new/worn-out)?
- What code of aircraft can use the runway – will it need modification?
- What is the safety record of the existing runway?
- Does the runway align to the predominant wind alignment and for how long?
- Does the runway create adverse noise problems for current and future populations and environmentally sensitive areas?
- Does the location of the runway create an effective operating model for airline operators?
- Does the location of the runway present any adverse problems for airport security?
- Are there any adverse operational issues with the existing runway?
- Does the existing runway have the correct rapid exit taxiway (RET)/ rapid access taxiway (RAT) and taxiway infrastructure? Can this be provided?

With the knowledge gained from answering at least the above questions on, say, the existing runway, the next step will be to understand if a new runway offers a better solution. This clause clearly demonstrates that an existing runway may be an asset that is a 'fixed constraint' but the benefits of fixing this asset need to be understood fully when compared with the more capital-investment-hungry solution of a new build.

In essence, the outline IATA masterplanning sequence is summarised to be:

Step 1. Understand the forecasted final design year capacity demand for the airport:

- peak aircraft movements
- peak passenger and baggage movements
- peak modal split ground traffic movements.

Aside. In Chapter 1 the full briefing requirements of forecasts are explained. The forecasts booklet needs to be managed and updated regularly and communicated effectively to the designers, as this sets the magnitude of the development and is the most important first step.

Step 2. Survey the airport site:

- obtain geographical data
- obtain geological data
- obtain meteorological data
- obtain environmental data.

Aside. The data identified above need to be provided to enable the designers to make informed decisions based on hard facts and knowledge. Often the temptation is to begin apron and runway design work before these data have been fully collated. That is fine but the development director runs the risk of incurring abortive costs, as it will be evident that data-led constraints will prove expensive or even impossible to resolve. This can be the case particularly with geological constraints. The key is to keep the design investigation work associated with later steps at a high level until step 2 data have been established.

Step 3. Understand the runway configuration options listed below and select the configuration that suits your airport development once a comprehensive analysis of the pros and cons have been examined:

- Runway configuration 1 – single runway.
- Runway configuration 2 – open V-runways.
- Runway configuration 3 – intersecting runways.
- Runway configuration 4 – staggered runways.
- Runway configuration 5 – dual parallel runways.
- Runway configuration 6 – multiple parallel runways.

The runway configuration should be aligned to best match the aircraft type and movement requirements, air traffic control (ATC) capability, geological limitations and meteorological conditions and environmental sensitivities.

Aside. This step can be extremely expensive to resolve and considerable ATC and ground movement modelling, runway length analysis, runway altitude and visibility analysis will be needed to establish the performance of the various runway configurations being considered. For airports that already exist the runway position will be heavily influenced by the availability of land and the suitability of the options relative to environmental concerns. New airports will have similar issues, but strategic land use and to some extent land availability is usually a step resolved in part by government departments before the team even gets to the design office.

Step 4. Set runway alignment. Align the proposed runway(s) to coincide with the prevailing wind directions.

Aside. Historic meteorological data can vary in quality depending normally on the country. The reliability of forecasted wind direction modelling is difficult to predict when the data are of good quality. Where these data are not readily available, the instances of existing runway experiences of cross winds will need to be examined carefully. This step 4 needs to be completed in parallel with the previous step 3 in order to strike a balance between achieving theoretical capacity versus runway usability.

Step 5. Apron planning. Determine and locate the number of aircraft stands required and the stand type (remote or gate serviced) needed to meet the service standard.

Aside. The service standards need to be examined fully. The walking distances for the passengers from the terminal to the gate need to be understood, as do the performance requirements for the aircraft from exit and entry to runway and the taxiing times. This is an area where computer modelling can be extremely valuable. The modelling of the taxiway and runways and the ATC should be completely linked. In this way the ideal time spent on the ground and aircraft ground movement congestion mitigation can all be examined. This step is often modelled in conjunction with the later step 6. Passenger experience will heavily dictate the optimum performance of this step.

Step 6. Taxiway planning. Provide the correct configuration and quantity of taxiways, ensuring that the runway(s) and stands are serviced adequately, with due consideration to the dynamics and potential congestion issues of the aircraft on the apron.

Aside. The taxiway position, number of holding points, the number and position of RETs and RATs will need to be modelled to determine the ideal capacity provision. Overall journey time and service standards and aircraft fuel burn while taxiing will be big issues to optimise.

Step 7. Terminal building definition. Size and position the ultimate terminal building(s), pier(s) and control tower within the appropriate development zones. There are primary and secondary development zones, which should be considered accordingly. Primary zones reside closest to the apron and existing terminal infrastructure and the key security points. Secondary development zones usually reside on the fringes of the airport campus and can be considered to be less operationally sensitive or valuable, though all real estate within the boundary of the airport should be considered to be coveted.

Aside. Chapter 3 of this book outlines the processes and the configurations that should be understood by the designers. The designers should consider the number of building levels, process options and the constraints imposed on the building.

Step 8. Terminal and apron positioning. Align the ultimate terminal building and piers to service the aircraft stands accordingly. Position fire services within the apron complex appropriately.

Step 9. Support processes and infrastructure. Size and position airport support processes such as (but not limited to) rail, bus, coach and passenger car access and parking facilities.

Step 10. Cargo and fuelling requirements. Position cargo and separate express facilities terminal and stands, aircraft maintenance hangars as required within the surplus development zone(s).

While each of the summarised IATA steps is simple and straightforward to understand in principle, the reality of most masterplan developments shall be that all aspects of the airport masterplan design steps will need to be developed in parallel simultaneously. The steps outlined in the IATA recommended practice are actually best referred to as a prioritisation sequence, e.g. step 6 has priority over step 7 and similarly step 7 has priority over step 8. There are always going to be exceptions to this rule where it can be proven that the operational issues or financial issues mean that the ideal solution for a later step means than an earlier step may not be optimised. The full convergence of design optimisation for each step is very difficult to achieve and it is likely that compromises will need to be made, especially when there are changes in forecast and long-term traffic volumes. In clause 2.2 that follows the compared analysis technique is explained. This technique allows all aspects of the design to be compared against one another and true comparative assessments carried out.

There is likely to be a design team dedicated to terminal building definition (step 7) and another team dedicated to terminal and apron positioning (step 8). In practice both teams will be working simultaneously but within agreed constraints. For example, both teams will have building footprint and height constraints agreed initially using benchmark comparative sizes. This will allow both teams to develop appropriate options in relative isolation until the latter location step 8 needs to be formally agreed.

2.2 Masterplan evaluation techniques (evaluation criteria/pairwise/weighting)

The analysis of masterplans or indeed detail designs needs to be rigorous. It is for this reason that airport planners should have an option evaluation tool/method that will allow them to assess subjectively the pros and cons of the various solutions being considered. If the airport planner puts forward a solution that is not fully thought through then the airport authority could be very vulnerable to criticism. If the objectors to the airport development can place doubt into the mind of the decision-making planning authority or discredit the integrity of the option evaluation team methods, then planning permission could be severely delayed and an irrecoverable situation produced.

While the technique described in this clause allows a fully comprehensive analysis using appropriate project drivers, it is important to note that the correct representation of all the stakeholders (airport/public) needs to be present during the evaluation scoring process. The present author has on many occasions used this technique during airport planning courses and split the class into two sections, with little regard for the skill base of pupils in each section. The conclusion found when doing this is that people rarely think outside of their area of responsibility/technical knowledge; i.e. if a group using this technique is predominantly of a financial background it is more likely that the outcome of the preferred option will be focused around financial issues. Clause 2.3 defines some of the more likely project drivers that can be used to assess the masterplans and detail design solutions. It is important that the skill base for evaluation has representatives from the passenger user groups, construction and cost integration expertise, environmental experts and airport and airline operational teams and lawyer representatives.

The pairwise analysis technique table (Fig. 2.1) allows the designer to compare each project driver against the others by scoring its relative importance and then to summarise the results to show a true representation of the preferences in terms of importance. It is important to be able to defend the decisions made. For example, in Fig. 2.1 driver A is slightly more important than driver B, driver C is slightly more important than driver A and driver D is significantly more important than all other drivers. The reasoning behind these decisions needs to be recorded for legal reasons. Lawyers should work with planners to ensure that designs are checked thoroughly. Any flaws in decisions will be highlighted by the objectors' lawyers, which can cause expensive delays.

Figure 2.2 defines a blank pairwise analysis technique table with the primary project drivers defined ready for use. This table can be expanded to list all of the issues to be evaluated.

Driver A is slightly more important than driver B

Driver C is slightly more important than driver A

Paired Analysis Comparison

	A	B	C	D	E	F	G	Total	
A		A1	C1	D3	A3	A0	A3	**7**	
B			C1	D2	B3	B0	B3	**6**	
C				D2	C3	C0	C3	**8**	
D					D5	F2	D3	**15**	33%
E						F3	G2	**0**	
F							F2	**7**	
G								**2**	

Driver D is significantly more important than all other drivers

2.1 A pairwise analysis technique table. Note that drivers should be determined by the airport planning team and could change from airport to airport. Scores should be determined locally using expert teams (5 = best, 0 = worst).

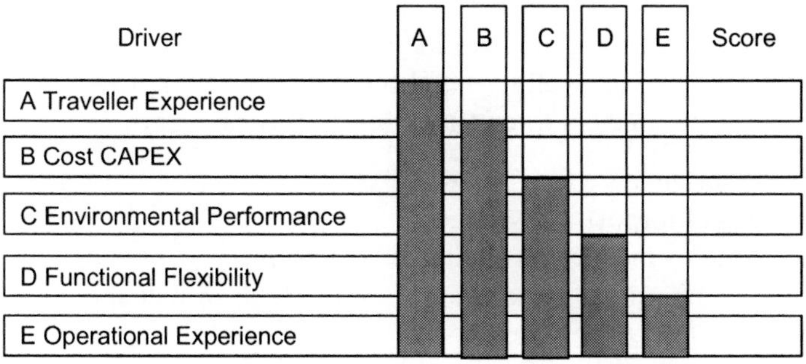

Driver	A	B	C	D	E	Score
A Traveller Experience						
B Cost CAPEX						
C Environmental Performance						
D Functional Flexibility						
E Operational Experience						

2.2 A blank pairwise analysis technique table with the primary project drivers defined ready for use.

2.3 Project drivers

When planning a terminal building there are many major decisions that need to be made. The decisions will influence the form, construction techniques, operational limitations, the economics and the overall performance of the airport. These can be typically categorised into the following project drivers, which include:

- passenger experience performance
- flexibility performance
- operational performance
- financial performance
- environmental performance
- airport building life

The airport terminal building will usually have a life of up to 40 years with major internal components, such as the baggage handling system and IT systems, needing replacement within this lifespan. It is for this reason that the terminal building should have a high degree of flexibility in each of the above categories listed.

2.3.1 Passenger experience performance

On the basis that a good passenger experience (a) will produce good results within passenger experience surveys and (b) will be likely to result in the airport gaining a good reputation and increased passenger traffic, it is therefore essential that high scores should be attained within these airport surveys. The types of issues that should be examined within this category will change according to the use of the terminal building. For example, a major transfer hub will have a set of criteria clearly linked to transfer connection performance while a low-cost terminal where these types of carriers do not typically operate airside transfer functions will have a different set of key performance indicators by which to be measured. The following criteria are generic to all terminal buildings and do not represent an exhaustive list of parameters to consider within this category:

- Passage to the terminal
  - Availability of airport rail services.
  - Journey time to the airport should be reasonable.
  - Journey time from car park/rail station to check-in hall should be minimal.
- Passage within the terminal
  - Check-in, security and passport control wait times should be reasonable.
  - Number of level changes should be minimal.
  - Good terminal layout minimises the need for wayfinding signage.
  - Walking distances should be reasonable, using people-moving systems if necessary.
  - Adequate number of clean toilets properly located throughout the passenger terminal.
  - Good retail shopping is expected by passengers and provides the airport authority with increased non-aeronautical revenues.
  - Sufficient number of restaurants offering a wide variety of food with different price levels.
  - Airline lounges are an important facility to business class passengers.
  - Baggage delivery wait times on arrival should be minimal.

- Passage from terminal to aircraft
  - Use of passenger boarding bridges provides an improved level of passenger service.
  - Use of remote stands with busing to meet peak traffic demands.
  - Walk on to the apron for low-cost carrier (LCC) operations.

2.3.2 Flexibility experience

Often this parameter is overlooked in so much as the immediate forecasts for the airport are considered to be fact, which is often not borne out in the medium to longer term. Also traffic characteristics will change over time with new airlines operating at the airport, the introduction of new aircraft types, new airline services and changes in government regulations. It is much wiser to have a passenger terminal building that is flexible and adaptable to different uses, e.g. that allows for the simple changing of internal partitions. The following criteria are generic to all terminal buildings and do not represent an exhaustive list of parameters to consider within this category:

- **Forecast flexibility.** The ability to adapt to changes in the forecast of traffic mix (e.g. short haul versus long haul, and the ability to respond positively to lower or higher forecast MPPA/BHR flows).
- **Phasing flexibility.** The ability to change the construction phasing delivery plan (e.g. allow for CAPEX changes with traffic up-turns or down-turns).
- **Flexibility to change the operational use of the facility.** The ability to change uses within the passenger terminal (e.g. allow for retail space to change to security search space, etc.).
- **Change in master plan philosophy.** The ability to be located in a zone that is of strategic importance (e.g. allow for additional runway required beyond the design horizon of the original master plan).

2.3.3 Operational performance

Good operational performance of the terminal building will be important to the passengers and staff who will use the passenger terminal. A well-thought-out terminal location and good passenger terminal design will enable staff to do their jobs in the most efficient way. The terminal building should exhibit characteristics that allow staff to function, to communicate and to operate safely within all areas of the terminal building, from front line check-in operations to back-of-house baggage hall operations. It is good practice to document and map out all of the processes that staff and passengers will undertake within the building. This process mapping can then be applied to the terminal concepts to see the effectiveness of the design. It is essential to use the knowledge of stakeholders to help evaluate the designs by getting representatives of the airlines, government agencies, retail, terminal operations and passenger user forums to help evaluate the early plans for the passenger terminal. While these evaluations might not change the form of the building proposed, often the knowledge presented can make the difference between a good design and an excellent design that is able to function effectively. A good example is the positioning of wayfinding signage. Operational staff should be consulted at the first opportunity to understand the characteristics of the users and the key signage interfaces and the displays that should be presented. The wayfinding strategy should be one of the first principles of the design to be fixed and certainly before the architectural design has been fixed.

The location of the passenger terminal should permit passengers to connect with aircraft in a reasonable period of time. The following criteria are generic to all terminal buildings and do not represent an exhaustive list of parameters to consider within this category:

- Effective and efficient staff operations
  - adequate space provisions for primary processes
  - easy vertical circulation for passenger, staff and trolley routes/flows
  - effective evacuation routes
  - facilities with good sight lines for staff
  - good natural lighting (check-in/ticketing/security search operations)
  - facilities that have good communication systems
  - facilities that have good security characteristics.
- Baggage operational efficiency
  - use of centralised baggage handling system with either manual or automatic sortation systems
  - use of centralised and efficiently linked hold baggage screening equipment
  - short distance from baggage hall to aircraft stands.
- Minimise turnaround times
  - terminals that are located sufficiently close to connecting infrastructure to enable fast passenger aircraft turnarounds (transfer distances kept short and, aircraft stands are located in effective positions with respect to taxiways and baggage halls).
- Redundancy
  - resilience of terminal location and design to mechanical breakdowns, accidents, incidents and acts of terrorism.

Table 2.1 IRR ranges for terminal buildings, piers and satellites

IRR percentage	Return observation	Commentary
7%–10%	Low IRR for international/domestic passenger terminal buildings	Witnessed when construction costs are high while retail and landing fees are not optimised High operating expenditure can lead to lower than expected IRR
	Medium IRR for stand-alone pier and satellite developments	Limited retail offers in piers and satellites means development revenues are limited to landing fees, limited retail and airline lounges
11%–15%	Average IRR – good for international/domestic passenger terminal buildings	Witnessed when good balance of retail to capacity provision Operating expenditure is realistic and befitting the operational environment
	High IRR for stand-alone pier and satellite developments	Significant retail within piers with high financial yield. At least two airline lounges present. Good landing fee revenue
+ 16%	High IRR – very good for international/domestic passenger terminal buildings	Very efficient terminal building. High capacity potential Operating costs are low owing to effective mechanical and electrical system designs and general environmental performance

2.3.4 Financial performance

The financial performance of the terminal building is all important as the building must be able to demonstrate a balance between cost, structural form, capacity provision and the internal rate of return (IRR) for the investment over a realistic period. The financial viability of terminal buildings should be the same as any major capital building investment. The IRR ranges for terminal buildings, piers and satellites are listed in Table 2.1. The IRR obtained for terminal buildings should be no less than 11%. For stand-alone piers and satellites the IRR should not be less than 7%.

It should be noted that domestic passenger terminals will typically return 25% less retail revenue than comparable international passenger terminals. This is due to reduced duty-free opportunities and market type, i.e. frequent flyer domestic business travellers usually spend less.

The following criteria are generic to all terminal buildings and do not represent an exhaustive list of parameters to consider within this category:

- capital cost of construction and land purchase/lease
- speed of building erection and time for full operational status
- cost of construction logistics, e.g. landfill versus excavation balance costs, etc.
- availability and magnitude of revenue streams
- minimised operational cost per passenger processed
- whole life cost analysis (asset replacement frequency).

2.3.5 Environmental performance

The environmental performance of the terminal building should be considered in two parts.

Part 1. Environmental impact to consider during design and operation
The design of the structure will need to consider the following items (this is not an exhaustive list):

- strategic views of the terminal site from local inhabitant areas (e.g. villages, hill tops, areas of natural beauty, etc.)
- river diversions – impact on ecology and bionetwork
- use of energy efficient technologies (e.g. ground source heat pump) and energy control processes
- road diversions
- landfill versus excavation balance assessment
- protected landmasses/species of wildlife issues
- light pollution – terminal design – glass box terminals can sometimes have poor light pollution characteristics
- light pollution – terminal control system – control systems can be used to limit impact to environment
- fuel supply routes to the terminal roads/gas pipelines/power cables
- CO_2 emissions from heating, ventilation and cooling systems
- waste water disposal from within the terminal
- de-icing fluid disposal where terminals have contact stands adjacent to the terminal.

Part 2. Environmental performance during construction
Projects can exhibit considerable impact to the environment during the construction sequence if not planned carefully. The following items should be considered in this regard:

- construction road access and congestion issues
- noise of construction close to site in prefabrication areas

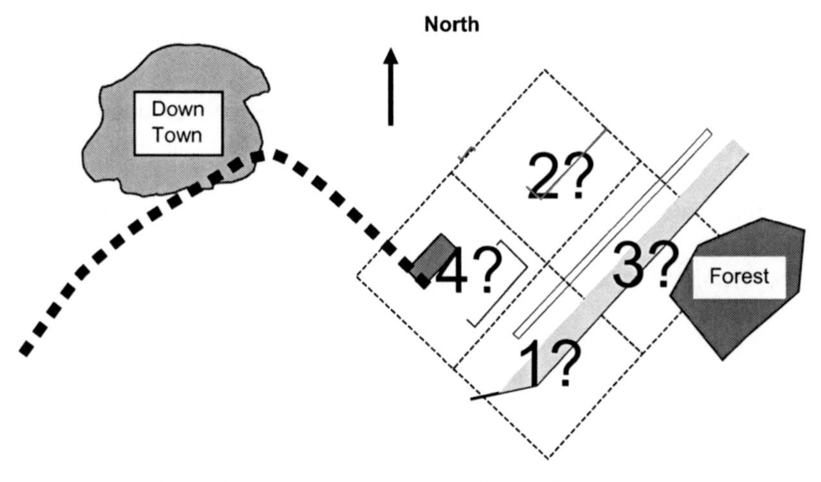

2.3 Terminal 1 and future terminal 2 positioning.

- night works (light and noise pollution)
- landfill and spoil disposal issues
- impact to air traffic control (ATC) and aircraft operations.

2.4 Airport case study

Figure 2.3 identifies an airport development with some fundamental issues that need to be addressed and accounted for when deciding the most appropriate location for the second terminal building. In reality the list of variables, needs and wants could be extensive and much more than listed in the case study example. The objective of this case study test is to propose a scenario of development and list some major development needs that need to be factored into the choice for the best location of the second terminal.

2.4.1 The scenario

The following airport is located on the mainland. The forecasted growth for the airport dictates that the airport needs to be developed and expanded to cope with an increase in traffic from 30 MPPa to 60 MPPa. The location of the eastern second runway has been decided following extensive ATC and noise mitigation evaluation. The location of the terminal is yet to be decided but it has been concluded that only one further terminal should be provided. The following information has been provided to help the reader to make an initial high-level decision on where best to locate the second terminal. On the airport masterplan given in Fig. 2.3, mark in the four boxes provided where it is best to locate the terminal, where 1 = best location and 4 = worst location.

2.4.2 Background characteristics of the airport location

- The town inhabitants recognise the need for expanded capacity and generally have accepted the proposed location for the new eastern runway.
- On the eastern side of airport there is an environmentally sensitive woodland.
- The airport processes point to point traffic but in the future is looking to become a transfer hub, so fast connections between terminals are important to business success.
- The southern side of the airport is vulnerable and security teams have noticed patrolling vehicles. It is regarded as being vulnerable to security problems.
- Road expansion is expensive and the airport does not have vast funds to extend road infrastructure.

It is important that, before this exercise is completed, an evaluation of the primary drivers is undertaken. For example, it may be more important to safeguard against security problems over environmental problems or airport performance.

2.4.3 Answer

The best location for this exercise is to locate the new airport terminal building in cell location 4. This promotes high connection performance both on the airport and for baggage operations; this location also has good security characteristics and avoids the environmentally sensitive woodland. The second best is cell location 2, but would experience less efficient transfer operations.

3
Airport terminal and pier/satellite planning

Abstract: This chapter reviews airport terminal and pier/satellite planning. It considers issues such as passenger segregation and flow through terminals and piers. It also discusses dwell periods in the context of the provision of landside and airside retail facilities and their contribution to financial performance.

Key words: airport terminal planning, satellite/pier planning, passenger segregation, passenger flow, landside retail, airside retail, dwell periods.

3.1 Passenger segregation

The legal requirements for passenger segregation within the terminal and piers and satellites can change considerably from state to state. The ICAO Annex 17 Security document statements (see Table 3.1) are intentionally not precise and are open to interpretation by the nations concerned to allow suitable flexibility.

3.1.1 The Schengen Agreement

States subscribing to the Schengen Agreement also have special dispensations, which allow far greater flexibility. The Schengen Agreement was originally created independently of the European Union, in part because of the lack of consensus amongst EU members and in part because those ready to implement the idea did not wish to wait for others to be ready to join. The Schengen Agreement was signed in Schengen, Luxembourg, on 15 June 1985, by Germany, France, Belgium, Netherlands and Luxembourg. Since Belgium, Netherlands and Luxembourg already had passport-free travel, the border tri-point of Luxembourg, Germany and France was considered a suitable place. The Convention implementing the Schengen Agreement, signed on 19 June 1990 by the five countries in Schengen, put the agreement into practice.

Table 3.1 **ICAO Annex 17 Clause 4.7, Measures Relating to Access Control (note that Clause 4.7.3 is highlighted for emphasis; ICAO material listed here is not version-controlled text; reproduced with kind permission from ICAO)**

4.7 Measures relating to access control

4.7.1 Each Contracting State shall ensure that security restricted areas are established at each airport serving international civil aviation and that procedures and identification systems are implemented in respect of persons and vehicles.

4.7.2 Each Contracting State shall ensure that appropriate security controls, including background checks on persons other than passengers granted unescorted access to security restricted areas of the airport, are implemented.

4.7.3 Each Contracting State shall require that measures are implemented to ensure adequate supervision over the movement of persons and vehicles to and from the aircraft in order to prevent unauthorized access to aircraft.

4.7.4 Recommendation.—Each Contracting State should ensure that identity documents issued to aircraft crew members conform to the relevant specifications set forth in Doc 9303, Machine Readable Travel Documents.

4.7.5 Recommendation.—Each Contracting State should ensure that persons other than passengers being granted access to security restricted areas, together with items carried, are screened at random in accordance with risk assessment carried out by the relevant national authorities.

4.7.6 Recommendation.—Each Contracting State should ensure that checks specified in 4.7.2 be reapplied on a regular basis to all persons granted unescorted access to security restricted areas.

All states that belong to the Schengen area are European Union members, except Norway, Iceland and Liechtenstein, which are members of the European Free Trade Association (EFTA). Switzerland joined the bloc's passport-free travel zone, the Schengen Area, from December 2008. Two EU members (the United Kingdom and Ireland) have opted not to participate fully in the Schengen system. The main reason that the non-EU states of Iceland and Norway joined was to preserve the Nordic Passport Union.

3.1.2 ICAO Annex 17 Security

ICAO Annex 17 Clause 4.7.3 (see Table 3.1) is probably the most relevant part of the document for passenger segregation. Designers, especially consultants working across national borders, should formally understand and record the relevant national interpretation of the acceptable conditions of mixing and separating departing, arriving and transferring passenger flows. It is the author's experience that designers and airport authorities will often claim to understand the legal requirements fully but actually do not fully appreciate the problems until they are noted in black and white.

Figure 3.1 provides a useful chart that can be used to record whether or

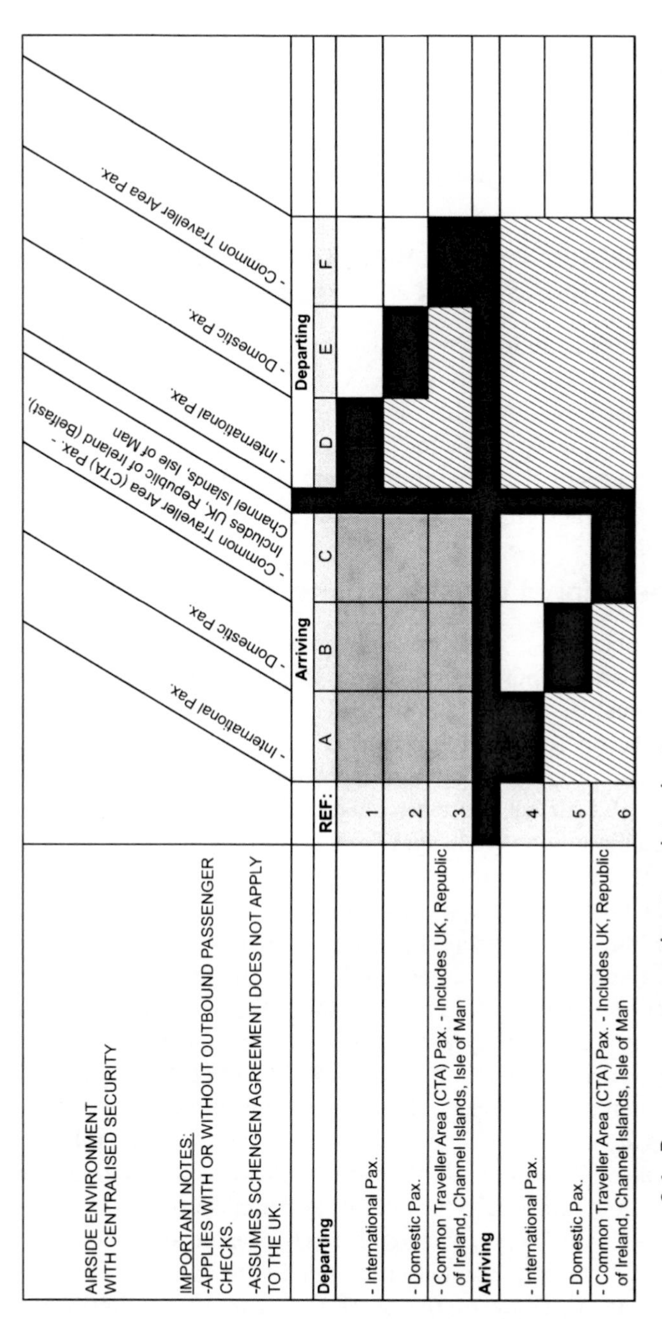

3.1 Passenger segregation template chart.

not departures, transfers and arriving passenger flows can mix or need to be segregated. The chart is shown with departures and arriving passenger flows but could equally be expanded to include transfer flows and special flows. When completing the chart it is important to have representatives from (a) national security policy (e.g. in the UK, the Department for Transport and Special Branch), (b) immigration services, (c) customs and excise, (d) airport operational management and (e) the design manager.

3.2 Passenger experience: process flow diagrams and interactive models

There are various techniques to assess what passenger flows might mean in practice for individual passengers. A two-dimensional approach is to construct a process-flow diagram. Figure 3.2 details the passenger processing map of a typical major terminal building. This can be combined with three-dimensional simulation. Figure 3.3 provides a snapshot of a commonly used walk-through simulation.

3.3 Operational considerations: airport/airlines

The primary objective of most privately owned airports is to support airline traffic growth and airport operator growth plans and to make sustainable profits. To enable this to be realised the airport operator needs to understand the operational requirements and growth plans of the airlines that operate or propose to operate from the airport. In high-level terms these airport and airline processes are basically the same the world over, although specific unique challenges from country to country and airline to airline are often evident.

Table 3.2 defines the common airport processes present within the airport terminal buildings. Passengers, operators and their staff are likely to call upon the building and its support infrastructure to provide the process functionality defined in Table 3.1 seamlessly. Technical teams should evaluate the process activity and map out the required building functionality. The processes defined in the table are defined at the very highest level. It is quite common as an example that the departures baggage flows need process map definition to be defined to the macro level, explaining how baggage is moved throughout the airport in fine detail, from passenger check-in to bag loading on to the aircraft and every process step in between. This can often run into tens of pages of step by step detail. All of this process step detail needs to be evaluated and refined with the airlines to maximise the operational efficiency of the airport. It can be a laborious task to

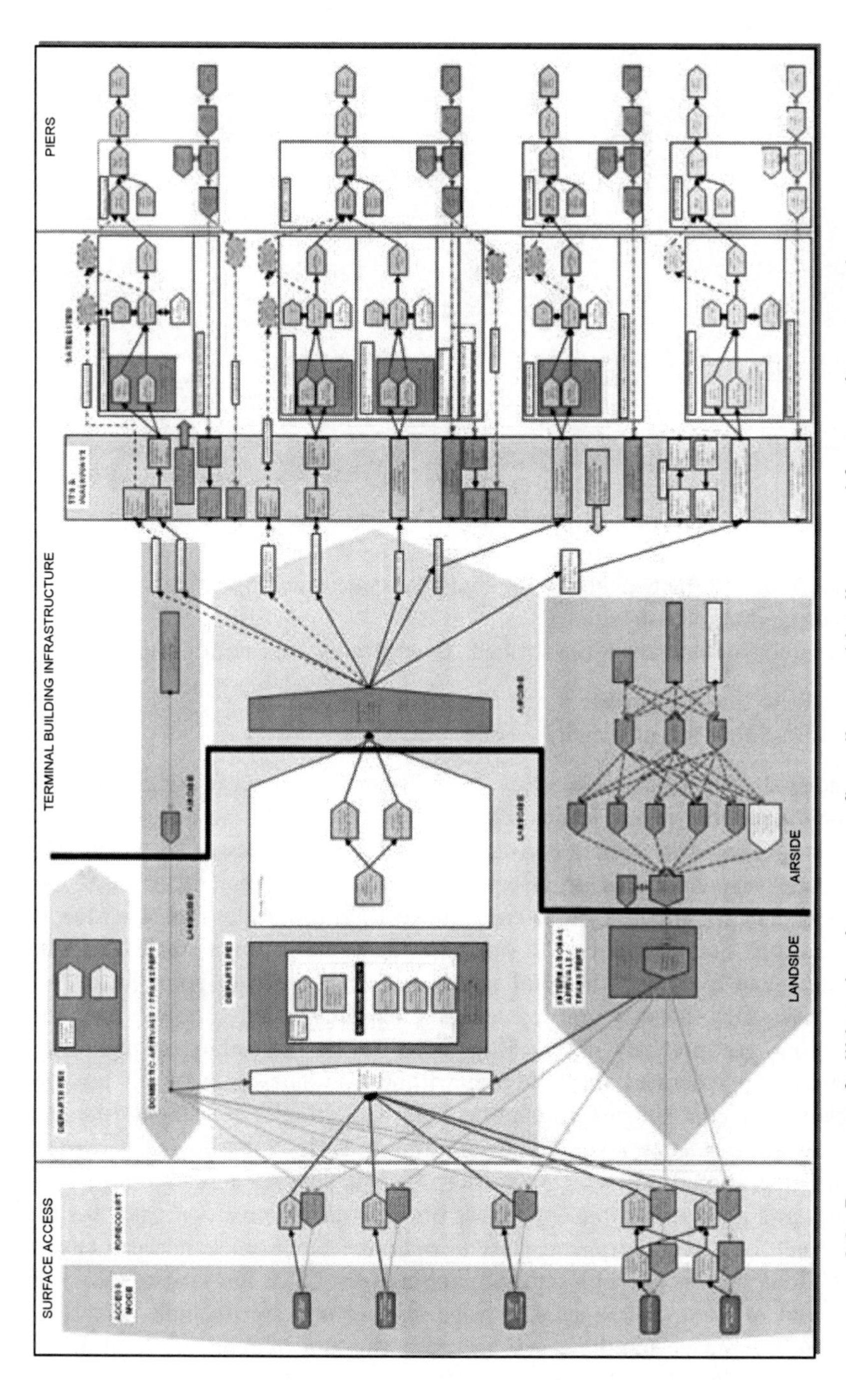

3.2 Passenger facilities planning process flow diagram (detail removed for security reasons).

3.3 Three-dimensional airport simulation: active walk through.

undertake but the benefits can be significant, as it can improve efficiency and security characteristics.

The airline operators are formed of essentially two main groups:

- those that are private limited companies (PLC) and
- those that are nationally owned.

National airlines have been often less efficient in their commercial approach when compared to the leaner, PLC airlines, especially the low-cost carrier airlines. Low-cost carriers constantly observe the processes by which they operate very closely in an effort to refine and optimise them. Staff and passengers are encouraged to conform to rigid timescales and windows of operation. The objective is to optimise turn-around times, maximise load factors and maximise the total number of daily point-to-point flights per aircraft. This ensures that passengers, baggage, staff, aircraft, fuel and catering are precisely where they need to be, according to the often demanding schedules they operate within. Revenues come from internet ticket sales, hold baggage surcharges, catering and sales of in-flight duty-free goods and, in some instances, even gaming cards. Some low-cost carriers limit operational costs by operating efficient aircraft, have reduced staff costs and do not use tour operators but instead obtain ticket sales via the internet. Low-cost carriers also try to operate from airports that can ensure high load factors but with reduced landing fees. Often low-cost carriers will opt out of using passenger air-bridge services and aim to limit the use of complex baggage handling systems with the aim of reducing landing fee charges to an absolute minimum. It is important to note that they all

Table 3.2 Operational airport processes commonly found within airport terminal buildings

Operational process	Functional objective
Baggage handling	Arrivals baggage flows Departures baggage flows Transfers baggage flows
Communication services	Post/radio/voice
Emergency management	Communication provision Emergency detection/alert Contingency planning Safety management Noise management
Ground transportation	Provision of public transport facilities Traffic control
Information provision	Information source Public address Visual information Creation of flight-related information Maintenance of flight-related information
Maintenance	Planned maintenance management Tools
People handling	Check-in Passenger movement Staff movement
Retail	Concession management Stores
Security	Pax access control Staff access control Vehicle access control Consumables and energy access control Baggage screening ID pass production Intruder detection Passenger screening Surveillance
Terminal management	Airline and handling agent liaison Passenger services Authority liaison Check-in desk allocation Operational management Trolley management
Apron management	Aircraft ground movement Aircraft handling, taxiway lighting /air/fuelling/power Runway safety apron lighting ice detection/friction testing/de-icing runway/apron cleaning and maintenance
Emergency management	Emergency response
Environmental management	Air quality analysis Air quality modelling and reporting Surface water quality

operate within and are required to comply with international/national regulations.

These two main groups are then further divided into:

- the scheduled airline (both legacy service and low-cost carriers)
- the charter operator.

Typically the charter operation will buy seats on selected scheduled airlines or operate on a few weekly flights to limited destinations. The scheduled airline will operate in multiple regions and has daily or hourly flight slot allocations. Some scheduled airlines sell groups of seats on selected flights as charter seats as a mechanism to guarantee base cost recovery.

3.4 Financial models

There are several financial models that can be applied to the operational revenue and costs of the airport operator. Airline groups and IATA member airlines generally prefer the 'single-till' financial approach. The concept of the 'single-till' is that the cost foundation for charges is based upon the cost of the airport facilities and services provided, net of contributions from non-aeronautical revenue sources. Under the 'single-till' or 'global residual' approach to rate setting, income streams from car parking and retail have the effect of lowering airport charges to airlines, while the airlines, in turn, assume the financial risk. Airport operators have the opposing view that the single-till approach subsidises the airlines and effectively puts adverse pressure on airports, particularly during periods of capacity constraint. It also creates an environment where operators do not have an incentive to develop new sources of non-aeronautical revenue.

Irrespective of how the financial model for the airport operates, the revenue generated from the retail space is of critical importance. When evaluating the 'value of a retail development' the following financial assessment techniques should be considered:

- internal rate of return (IRR)
- net present value
- the weighted average cost of capital (WACC).

Table 2.1 in Chapter 2 defines the typical levels of rate of return that can be expected from new terminal, contact pier and satellite structures that include retail facilities. If these structures already exist, then obviously the capital cost of the project decreases and the IRR should become significantly better than defined within Table 2.1. The rates of return identified should be achievable if construction costs are controlled appropriately and careful calculation of revenue expectation is made.

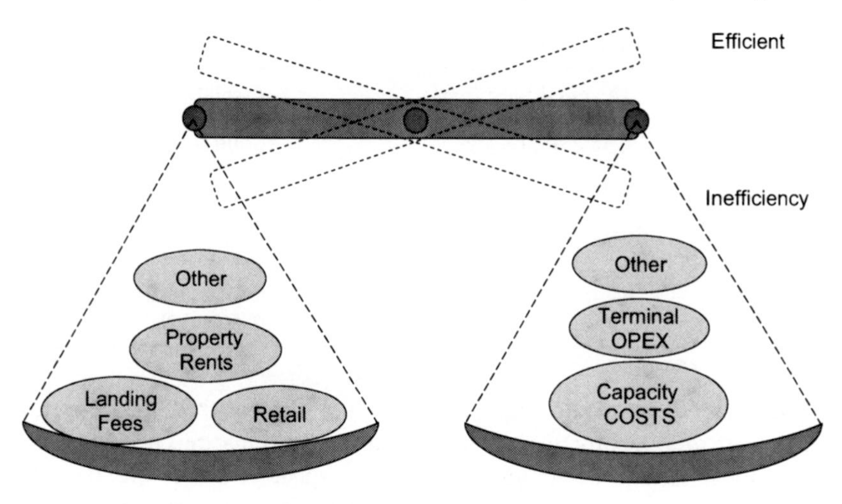

3.4 Passenger facilities planning.

The IRR will be affected by:

- the capital cost to provide the facility
- the taxation regime applied locally
- the operating costs
- the incremental revenue generated from the retail space
- the design life of the retail product.

It is important to note that airports also rely on income streams from airport hotels and associated business parks, which can factor into the single-till and subsidise aeronautical fees. These opportunities should be fully understood.

Figure 3.4 shows that the financial performance of a terminal is a dynamic balance between the provision of efficient terminal capacity and optimised revenue streams. If the airport designers can get the balance right, then it is possible to generate an efficient and attractive airport.

3.5 Retail design principles

The modern passenger terminal building must incorporate various passenger support and commercial facilities. These are essential to provide passengers with the levels of service and convenience they expect. The revenue from retail and food and beverage outlets, plus potentially airline lounges, should generate sufficient non-aeronautical revenues to allow the airport operator to keep aeronautical charges to the airlines at a reasonable level.

Table 3.3 identifies the various configurations of departures lounges that can be found, along with the recommendations to be used in each case. The definitions of terminal type are:

Table 3.3 Terminal retail models

Item	Terminal type	Available retail solution
1	Architectural	Central hub High street IATA model
2	Box terminal	Central hub IATA model
3	Multibox terminal	Central hub IATA model

- **Architectural terminal.** This is a terminal design that incorporates signature architecture, usually incorporating combinations of linear and non-linear structural design features. The space is essentially a single enclosure and can be on a single level or on multiple levels. Examples include Hong Kong Airport and Heathrow Terminal 5.
- **Box terminal.** This is a terminal design that incorporates linear architecture only and could be developed by signature architects or local architects. The space generated is essentially a single box and could be either a single-level or a multilevel arrangement. Examples include Stansted Airport(s) and Gatwick North Terminal.
- **Multibox terminal.** The multibox is a newer design approach which uses the connection of multiple box enclosures each with a discrete space sized to accommodate one function, e.g. box 1 = check-in, box 2 = security, etc. The theory is that each space can be expanded to accommodate the increase in capacity needed independently from the adjoining process block. There are positives and negatives associated with this model, which need to be fully examined. Examples include many minor Category A airports processing LCCs within Europe.

Definitions of available retail solutions are:

- **Central hub.** This is where passengers enter a common lounge after outbound security and immigration and provides centralised airside retail, seating, toilets and circulation space.
- **High street.** This is where the passengers flow through a relatively narrow corridor en route to the gate, with retail positioned either side of the corridor or just on one side, emulating the non-airport 'high street' retail experience.
- **IATA model.** This is as denoted in the IATA 9th edition *Airport Development Reference Manual,* Chapter J7, Fig. J7.1. The principle of the lounge is the same as the central hub previously described but with the difference that a direct clear through-route to connecting piers or satellites is provided. Some conclude that the effect of this option is to

reduce revenue from sales of goods in the retail lounge in favour, arguably, of a less complicated way through the terminal for passengers. This is very debatable.

The retail space solution that can be applied to a passenger terminal will be dependent upon the architectural space available, if it is an existing building. The central hub solution should be the recommended solution to adopt, if possible. If the building architecture is yet to be defined then again the central hub retail solution should be adopted, as this will yield the greatest income per passenger irrespective of terminal type. Figure 3.5 shows the central hub arrangement and the recommended position of the entrance to the departures retail unit. Upon entry into the departures lounge retail space, passengers should be able to immediately orient themselves, such as being able to see most of the shop frontage branding, have a clear sight of the flight information displays and easily find the exit points from the departures lounge. It is important to make the passenger feel in control of the journey to the gate. A relaxed passenger that is able to see the departing gate or know how long it will take to walk to the gate or connect to the gate will be the passenger who is more likely to take time to spend in retail.

The line of sight to the shops should be such that there are no obstructions to view key retail shop space. It is important that the airport operator understands the profile of the passengers that will be using the airport and then provides the type of retail shops that will meet the shopping needs of the passengers going through the departure lounge retail space. The departures lounge retail space should ideally be on one level. If the departure lounge is split on multiple levels then there should be no more than two levels and it will be important to use the primary retail space effectively on the main level. Secondary retail space, including food and beverage and toilet facilities, should be located on the upper or mezzanine level within the departures lounge. Airline lounges can also be located on the upper level close to the retail space. The primary retail level (entrance level) should accommodate prime retail shops and the seating should enable the passengers to see all the shops from the positions of the seats. The seating, the through-route corridors and the circulation space on the primary level should be configured such that the route to the exit(s) can be easily seen and ensures maximum exposure of the retail units.

The retail dwell time will effectively dictate the dimensions *A* and *B* denoted within Fig. 3.5. As the retail dwell decreases so will the ability of passengers to shop and get to the outer edges/perimeter of the retail space and allocate time spending on duty-free type purchases. Airports are, however, beginning to define themselves as destinations in their own right and so this space could become quite large. Examples of this include Terminal 5 Heathrow and Hong Kong Airport.

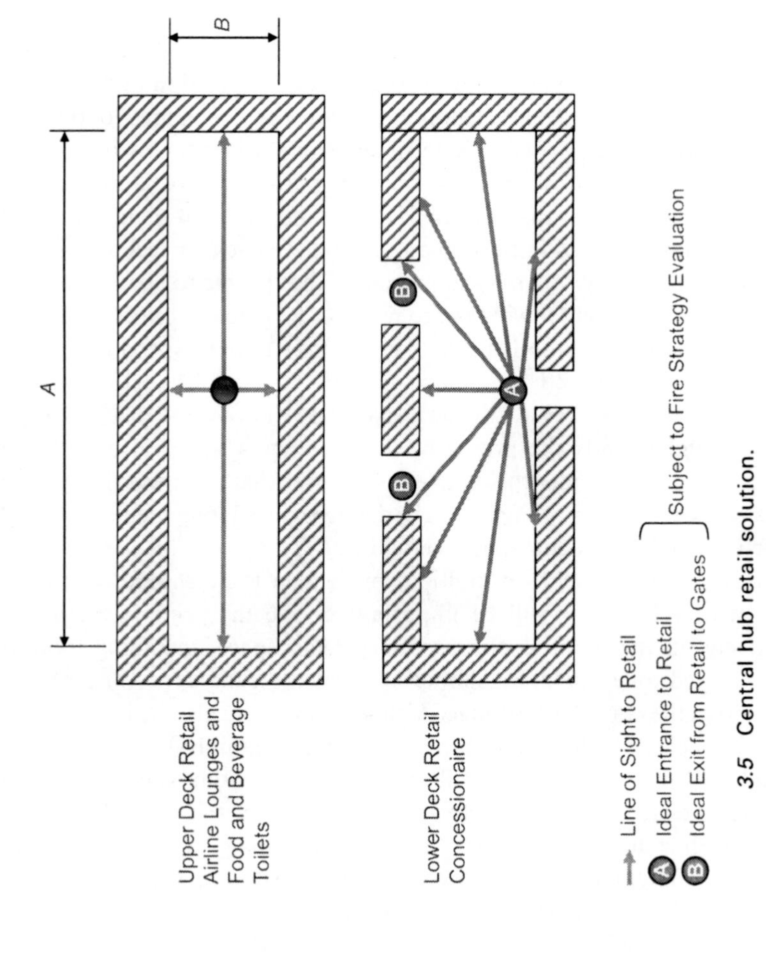

3.5 Central hub retail solution.

Table 3.4 Composite international departures lounge (IDL) items

Item	Description of space requirement	Building level/comments
1	Circulation space	All levels, particularly primary entrance level
2	Passenger seating	Primary entrance level only
3	Flight information display screens	All levels, particularly primary entrance level
4	Emergency signage and public address systems	All levels, particularly primary entrance level
5	Originating departing passenger flow input point	Primary entrance level only
6	Transfer departing passenger flow input point	Primary entrance level only particularly if there is significant transfer passenger flows
7	Passenger toilets	Primary or secondary passenger level; secondary preferred if retail area demand dictates space should be provided
8	Advertising signage	All levels
9	Disabled passenger processing facilities	All levels
10	Food and beverage	Primary or secondary passenger level; secondary preferred if retail area demand dictates space should be provided

It is recommended that, for each terminal departures lounge, the optimised/maximum dimensions, subject to busy hour rate demand are $A = 200$ m maximum by $B = 150$ m maximum. A common mistake made at many older airports is that the departures lounge is too narrow and as a result the retail space is long and narrow with insufficient retail shops to meet passengers' shopping requirements. This means that potential retail revenue is lost and the airport operator fails to generate additional non-aeronautical revenues. Table 3.4 lists items that should be considered for inclusion in the typical departures lounge.

3.6 Landside retail

Landside retail in most expanding airports is actually on the decline. Research has shown that it is operationally advantageous to limit landside retail to the absolute minimum. Landside retail should be targeted at 'meeter' and 'greeter' visitors to the airports and should not be targeted at the passengers. This will ensure that passengers are encouraged to progress through the security and emigration processes as fast as possible and not dwell in landside retail. This aligns well with optimised airside duty-free sales and timely aircraft embarkation. Where airports in the past have developed large landside retail spaces they can often be used by non-airport passengers. This creates the dilemma of using expensive airport capacity for purposes that do not promote increased true passenger throughput.

Figures 3.6 and 3.7 define the appropriate proportions and location of landside and airside retail. Where landside retail is required it can incorporate both (a) retail and (b) food concession facilities. Landside facilities should be sized to align with the volume of passengers, landside staff and 'meeter/greeters' using the facilities. The retail element of this facility will typically account for 20% of the total retail space provided within the terminal. Figure 3.8 illustrates examples of landside and airside retail facilities.

3.7 Dwell periods

Figure 3.9 highlights the following sensitivities associated with dwell periods within the departures lounge and details a typical passenger processing time line.

- The first passenger through check-in has more retail time opportunity and the last passenger through check-in has less retail time opportunity. Therefore the airport authority should allow the airlines to start the check-in process well ahead of the flight departure time.
- Slow check-in, security and outbound immigration reduces the dwell time opportunity in the retail shopping area. This is actually a difficult balance to achieve. The common goal should be to progress the passenger to the departure lounge for as long as practical, giving due consideration to travel time to the departure gate and gate processing time demands.
- Excessively early announcements asking passengers to proceed to the gate lounge reduce retail dwell opportunity. In practice airlines prefer to control this, particularly LCCs, which have a major requirement for a quick turn-around of aircraft in the minimum time and cannot afford late passengers or indeed the time taken to remove passengers' baggage from the hold for those passengers that do not arrive at the gate while the gate is open. It is recommended that airlines and the airport agree on gate announcement rules to ensure that call to gate announcements are not exploited either in favour of excessive airport operators' retail spending time or the ability of the airline to reasonably guarantee the presence of passengers at the gate in sufficient time. It is a fine rational balance that is required.
- The time the passenger spends within the departures lounge is proportional to the above points, plus it will be a reflection of the quality and diversity of the retail offering and obviously it is also

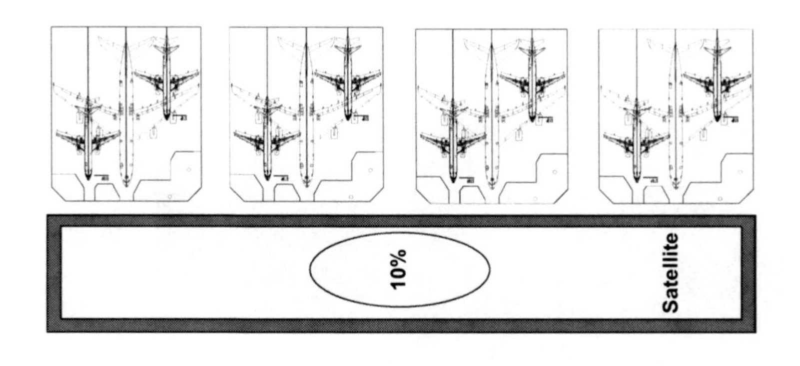

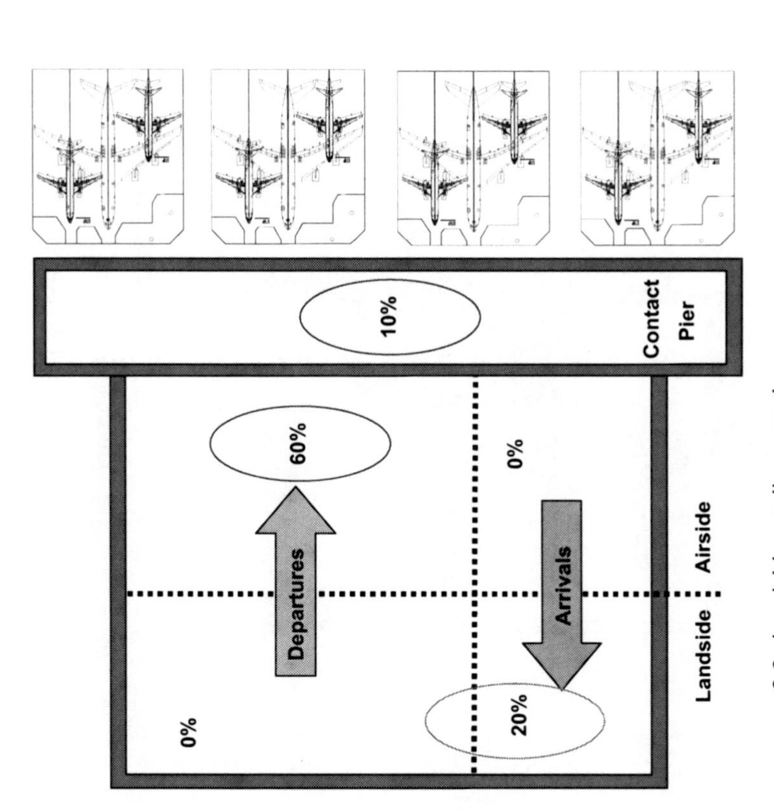

3.6 Landside retail proportions.

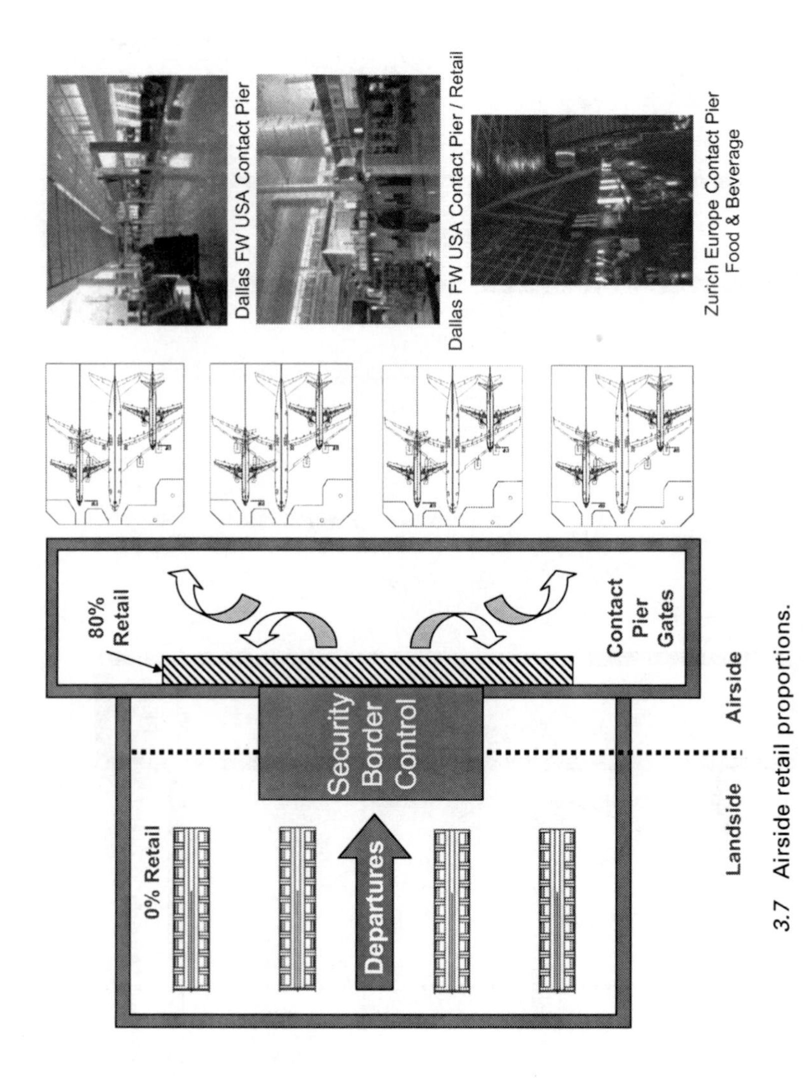

3.7 Airside retail proportions.

3.8 Examples of retail facilities: (a) landside; (b) and (c) airside (departure lounge).

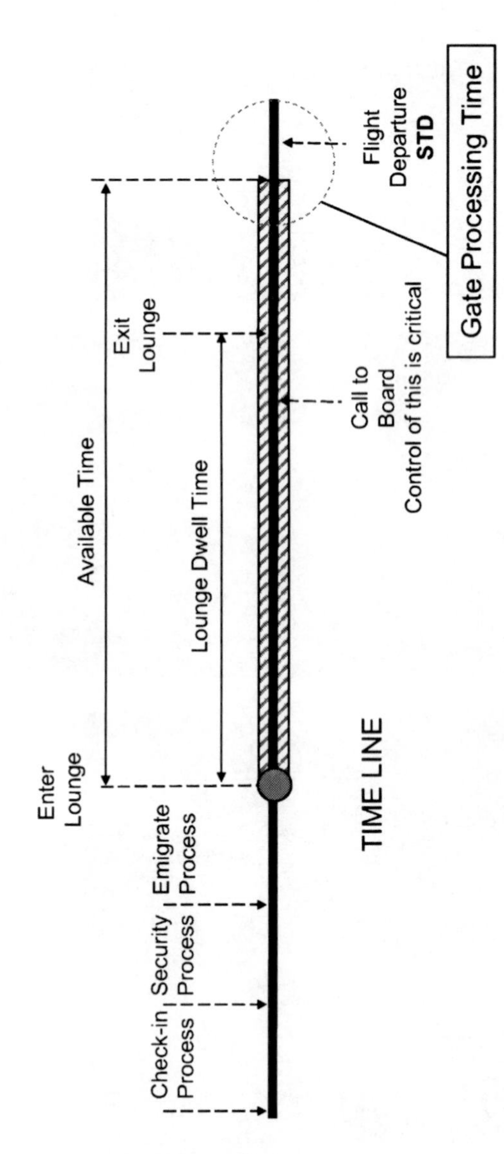

3.9 Dwell periods for a single flight.

associated with the disposable income of the passengers. Frequent travellers to an airport may spend different amounts of time in the departures lounge for each flight depending on their schedule and shopping requirements; one day the passenger may be in a hurry to catch the flight and pass quickly through the departures lounge, while on another day the passenger may have ample time to spend shopping or perhaps visit an airline lounge, or both.

It is important to develop a common departures lounge that can accommodate international, domestic and transfer passengers in shared and common space, subject to the local domestic laws. To achieve this it will be necessary to provide the downstream positive passenger identification processes to confirm that the correct passengers enter the correct domestic or international contact pier or satellite

3.8 Airside retail: departures lounge/satellite/pier

To understand the true functional requirements of the departures lounge, it is first necessary to understand the departing passenger profile for the check-in function and the speed and effectiveness of the security and outbound immigration functions. Figure 3.9 details a time line for a typical single flight. It should be noted that there would be multiple flight time lines overlapping, which creates the net busy hour rate. It can be seen that there is a clear objective on a flight-by-flight basis for passengers to be processed through check-in, security and outbound immigration, as quickly as possible, to allow passengers to spend the maximum time within the departures lounge before proceeding to the departure gate lounge and boarding the aircraft.

The majority of retail income will come from the retail space in the departures lounge. It is for this reason that careful planning of this space is required and should be made early in the planning process for the passenger terminal in order to maximise its revenue-generating potential. The space needed will be a function of the following factors:

- departures lounge busy hour rate
- passenger dwell time in retail area
- busy hour rate peak factor
- passengers in departure lounge
- MPPA processed through the airport
- percentage of seating area
- space per pax (m^2)
- seating area (m^2)
- retail area (m^2)
- percentage circulation space.

3.9 Planning the size of terminal buildings

The truly dynamic nature of passenger movements through the terminal building and its support infrastructure can be statically calculated for masterplanning purposes only by identifying key variables such as peak numbers of passengers entering the terminal, departing or arriving on flights. These can then be used to estimate key parameters such as the size of the departures concourse required. The following software is available to enable the core passenger processing and retail components of the internal space to be calculated statically: *Reference calculations for planning airport terminal buildings: a supplement to* ***The independent airport planning manual***. This software is available by contacting Woodhead Publishing Limited at: www.woodheadpublishing.com.

As the building design is taken to the next level of design from concept to feasibility, it will be necessary to employ the services of simulation models. These simulations allow the behaviour of passengers in real time to be understood. It is important to note that passenger movement simulations should be completed in conjunction with computational fluid dynamic (CFD) software. CFD software allows fire progression within a building to be analysed, which, if completed with passenger movement simulations, can be extremely useful in determining evacuation times and smoke occupation.

4

Airport baggage handling design

Abstract: This chapter discusses airport baggage handling systems design. It outlines different categories of baggage handling systems before going on to review user and legislative requirements. It describes different screening processes before discussing the advantages and disadvantages of various hold baggage screening (HBS) processes and locations.

Key words: airport baggage handling, hold baggage screening (HBS).

4.1 Categories of baggage handling systems

Historically airport baggage handling systems (BHSs) have been categorised by size and complexity according to the volume of baggage that they process per hour.

4.1.1 Category A baggage handling system

Category A baggage handling systems experience peak baggage flow rates of less than or equal to 999 bags/peak hour. It is the recommendation of IATA that the sortation systems are manual or automatic. Manual sortation racetracks can be used (subject to health and safety legislation constraints) or automatic sortation can be employed using conveyors with pusher devices or verti-sorter units. If the primary Category A baggage system fails, the airport system capability should exist to process a high proportion of the baggage through a redundancy system. For a Category A baggage handling system this should be at least a manual sortation process and baggage hall system, which is covered from the elements, safe to operate and secure and compliant with national and international ICAO mandates. The baggage hall or apron area used for this redundancy operation should be at least twice the size of the racetrack system and vehicle space normally provided plus appropriate airport operator staff sortation assistance during system downtime. Where an automatic system is provided, redundancy provision should be made for an automatic sortation system capable of processing 50% of isolated peak baggage flow rate at all times.

4.1.2 Category B baggage handling system

Category B baggage handling systems are used where the peak baggage flow is envisaged to be greater than 1000 bags/peak hour or less than 4999 bags/peak hour. IATA recommends the use of automatic baggage sortation devices only, when processing this demand range. The term 'automatic' describes baggage sortation device(s) such as conveyors with pushers or with verti-sorters, tilt tray sorters or basic destination coded vehicle trays. Crane and rack sortation and storage systems are not described by IATA in this category but it is reasonable to conclude that they could be used with Category B baggage handling systems, particularly if batch build loading is used. In the event of a failure, the baggage system should have the redundancy capability to be able to automatically sort 75% of peak baggage flow rate.

4.1.3 Category C baggage handling system

Category C baggage handling systems are used where the peak baggage flow rate is envisaged to be greater than 5000 bags/peak hour. IATA recommends that only automatic tilt tray sortation or top end destination coded vehicle tray sortation units are used when sorting these rates. Crane and rack sortation and storage systems are not described by IATA in this category but it is reasonable to conclude, as with the Category B baggage handling systems, that they could and almost certainly would be used within a typical Category C baggage handling system. The scale of the logistics of airports processing this volume and the complexity of the baggage route alternatives means that a Category C baggage handling system will be a highly complex control mechanism, which will need to be very flexible. When the primary Category C baggage handling system fails it should be possible to process no less than 75% of the peak baggage flow rate automatically at all times.

4.2 User requirements specification (URS)

A comprehensive baggage user requirements specification (URS) should be developed before the planning of the baggage handling system commences. The recommended heading types are listed below. It is commonly regarded that the baggage handling system is the heart of the terminal, functioning out of sight of the passengers with limited start of journey and end of journey passenger interfaces. The truth is that, owing to the complexities of multiple destinations, multiple applicable security screening mandates and massive changes in passenger arriving profiles, the airport baggage handling system needs to be sophisticated to deal with these scenarios. The baggage handling system must be well thought out using the correct technology and the correct operational processes. Without this the terminal as a whole will not function.

The URS should define baggage flow rates (peak hour and normal flow) for the specific parts of the baggage handling system to be supplied. The design life of the system should be defined. This will not always be the maximum possible/achievable for the technology. The design life period for most baggage handling systems, if well maintained, is typically 15 years. The airport planner should align the size of the baggage system and the quality of the baggage system to knowledge associated with the masterplan. It is wasteful to design a baggage system with a design life of 15 years if you know that your terminal in day 1 form is only going to reside in this location for 10 years and will need to be replaced and/or relocated. This basic thinking can save millions for an airport if the masterplan is well thought out with system life checks.

The URS contents must include:

- baggage system performance expectation (in system time)
- baggage input statement – what types of baggage sizes can be accommodated
- baggage system functionality statement – how the baggage system works in practice
- physical components of baggage handling system
- system availability (operational time and acceptable system downtime)
- baggage travel times aligned to minimum connection time (MCT)
- baggage make-up lengths and class segregations
- airline system interfaces (SITA, etc.)
- baggage reconciliation capability (AAA, etc.)
- baggage tug types and container types
- baggage tractor battery-charging facilities
- container storage facilities
- flight allocation systems and facilities
- processing of oversized baggage

4.3 Hold and hand baggage screening legislative requirements

There are essentially two international hold and hand baggage screening legislative requirements. ICAO Annex 17, Security, which is a truly globally targeted document and one of the ECAC that is targeted at European countries only. ECAC documentation makes reference to ICAO Annex 17 and therefore the ICAO document should be considered as the single most important legislative requirement. The ECAC document is very much more focused on the practicalities of implementing updated security measures. The author recommends that even non-European airports should seek to obtain a copy of the ECAC Document 30 (Security) for best practice reference. There is obviously also national legislation, which must be observed and complied with in each respective country. In the United Kingdom, for example, the Department for Transport (DfT) stipulates the legislative requirements by which airports should provide security, but this, like the Federal Aviation Authority security documentation, is a derivative and interpretation of the ICAO documentation. In summary ICAO Annex 17 is the common denominator security document and all other security documents support this overriding document, which is endorsed by all NATO member countries via the United Nations. The specific ICAO Annex 17 Security Clauses of interest are shown in Table 4.1.

The above references to ICAO clauses are obviously not controlled. The author recommends that the designer obtains the appropriate controlled version documents produced by ICAO and ECAC (if permitted) and IATA.

4.4 Current and future baggage screening processes

Arguably the complexity of the modern baggage system is a direct result of the separation of the bag from the passenger at check-in. Designers should challenge the current processes and look for innovative ways in which to process passengers and their baggage. There are a few reasons why hold baggage is removed from the passenger before the passenger goes through the restricted zone landside boundary. These include:

(a) to ensure that the airlines through use of the baggage handling system have the maximum time possible to screen and sort the baggage complying with Section 4.2 above;
(b) to ensure that the central search and immigration processes for passengers are straightforward and clearly defined and processing time is minimised;
(c) to ensure passengers entering the airside departures lounges are not caring for bulky hold baggage;
(d) to ensure that passengers are given the maximum possible time to orientate themselves, relax and, most importantly, have the optimum opportunity to use and spend money in the airport retail lounges.

If the terminal and baggage handling designers can create a process with supporting systems that will allow the passengers to retain their hold baggage and positively meet the objectives stated above, then the result could be a less complex baggage handling system. This new baggage process

Table 4.1 **ICAO Annex 17 Chapter 4, Security Clauses Relating to Baggage (ref. 1/7/02 4-2; ICAO material listed here is not version-controlled text; reproduced with kind permission from ICAO)**

4.3 Measures relating to passengers and their cabin baggage

4.3.1 Each Contracting State shall establish measures to ensure that originating passengers and their cabin baggage are screened prior to boarding an aircraft engaged in international civil aviation operations.

4.3.2 Each Contracting State shall ensure that transfer and transit passengers and their cabin baggage are subjected to adequate security controls to prevent unauthorized articles from being taken on board aircraft engaged in international civil aviation operations.

4.3.3 Each Contracting State shall ensure that there is no possibility of mixing or contact between passengers subjected to security control and other persons not subjected to such control after the security screening points at airports serving international civil aviation have been passed; if mixing or contact does take place, the passengers concerned and their cabin baggage shall be re-screened before boarding an aircraft.

4.4 Measures relating to hold baggage

4.4.1 Each Contracting State shall establish measures to ensure that hold baggage is subjected to appropriate security controls prior to being loaded into an aircraft engaged in international civil aviation operations.

4.4.2 Each Contracting State shall establish measures to ensure that hold baggage intended for carriage on passenger flights is protected from unauthorized interference from the point it is checked in, whether at an airport or elsewhere, until it is placed on board an aircraft.

4.4.3 Each Contracting State shall establish measures to ensure that operators when providing service from that State do not transport the baggage of passengers who are not on board the aircraft unless that baggage is subjected to appropriate security controls which may include screening.

4.4.4 Each Contracting State shall require the establishment of secure storage areas at airports serving international civil aviation, where mishandled baggage may be held until forwarded, claimed or disposed of in accordance with local laws.

4.4.5 Each Contracting State shall establish measures to ensure that consignments checked in as baggage by courier services for carriage on passenger aircraft engaged in international civil aviation operations are screened.

4.4.6 Each Contracting State shall ensure that transfer hold baggage is subjected to appropriate security controls to prevent unauthorized articles from being taken on board aircraft engaged in international civil aviation operations.

4.4.7 Each Contracting State shall establish measures to ensure that aircraft operators when providing a passenger service from that State transport only hold baggage which is authorized for carriage in accordance with the requirements specified in the national civil aviation security programme.

4.4.8 From 1 January 2006, each Contracting State shall establish measures to ensure that originating hold baggage intended to be carried in an aircraft engaged in international civil aviation operations is screened prior to being loaded into the aircraft.

4.4.9 Recommendation.— Each Contracting State should establish measures to ensure that originating hold baggage intended to be carried in an aircraft engaged in international civil aviation operations is screened prior to being loaded into the aircraft.

4.4.10 Recommendation.— Each Contracting State should take the necessary measures to ensure that unidentified baggage is placed in a protected and isolated area until such time as it is ascertained that it does not contain any explosives or other dangerous substances.

is likely to be operationally developed and used in the future. To date many airports have considered this – particularly smaller airports – but the strategic risks are quite considerable. So far there have been no major significant airport users willing to take up the use of this baggage processing approach. The major disadvantages are that:

(a) the airport operator will witness reduced retail sales from duty free;
(b) hold baggage enters the restricted zone with the passenger, albeit previously screened at this point, and the passenger can tamper with baggage and the bag tags;
(c) the airlines do not have the maximum time in which to sort the baggage and load the baggage into the aircraft;
(d) early bags are not easily accommodated.

An alternative possible future hold baggage processing sequence could include the following seven steps. It should be noted that these process sequence steps are not currently recommended for terminals processing in excess of 5 MPPA and generally are *not* recommended as best practice at this time. Importantly, with further development this process could become an effective way in which to process hold baggage with minimum baggage sortation system costs. A major advantage with the process steps 1 to 7 inclusive could include: (a) dramatically reduced baggage handling system costs; (b) effective bag to passenger reconciliation if the hold and hand baggage fails the screening process; (c) reduction in airline responsibility associated with mis-sorted baggage.

Step 1. Passengers enter the terminal.
Step 2. Passengers have hold baggage weighed and *all* baggage hold and handheld is labelled and a passenger boarding pass issued.
Step 3. Passenger takes hold and hand luggage to the Restricted Zone (RZ) boundary – boarding pass is checked. Refer to Chapter 1 for a definition of the restricted zone.
Step 4. Passenger places all baggage into baggage screening process (combined hold and hand baggage process). Cleared hold and hand baggage is returned to the passenger.

- Step 4a. Passengers are screened.
- Step 4b. Cleared passengers with cleared hold and hand baggage enter the baggage flight sort drop point zone. This is where the passenger is confronted with multiple flight drop points and self sorts the baggage to the correct flight bag drop position where a member of staff checks the bag tag.

Step 5. Passengers with only hand baggage present themselves to emigration checks according to national border requirements.
Step 6. Passengers with hand baggage enter the IDL, relax and shop.

Step 7. Passengers with hand luggage go to the gate and board aircraft.

Figure 4.1 compares typical passenger and baggage flows within (left image) a conventional terminal arrangement to one where the passenger moves the baggage through the main terminal building (right image) and baggage is dropped at the entrance of the contact pier, thereby simplifying baggage sortation needs considerably, albeit to the detriment of passenger service and ease of passenger movement through the terminal retail space.

4.5 Advances in baggage system automation

The conventional 'automatic' sortation airport baggage system process has remained the same for the last 30 years with only relatively minor advances being introduced. These historic changes have usually been attributed to:

- advances in computer controls;
- more reliable equipment;
- improved sortation capability;
- introduction of automatic baggage storage systems.

Future baggage systems will need to be much more efficient than current baggage systems, with greater flexibility and perhaps with greater functionality. This mix of requirements must translate into reduced airline and airport operating costs. This will be difficult to achieve, especially as innovation is costly and risky. Even the most automatic of existing baggage handling systems are essentially 'open loop' systems; i.e. at the front end of the baggage process they are heavily dependent on manual passenger and staff performance to ensure that the bags are injected into the baggage system effectively. At the tail 'output' end of the baggage process the bags are manually moved from the baggage system chutes, laterals or racetracks into the baggage trucks. It is this latter part of the total process where the main advances in technology can improve system time and could dramatically reduce injuries to staff who are required to move some heavy bags repetitively.

So where are the inefficiencies and risks with current baggage system processes? There are broadly two areas of main inefficiency and risk:

1. **Number of baggage handling operatives.** Baggage operatives are required to process bags while flights are open. If a flight is open for 2–3 hours then resources are required to deal with the load of baggage expected within that open period. The profile of load will vary according to the type of flight going through a baggage system. Short-haul point-to-point flights with a predominantly business sector of clients will have fewer bags and the passengers will generally turn up at the latest possible moment. However, long-haul flights with high transferring

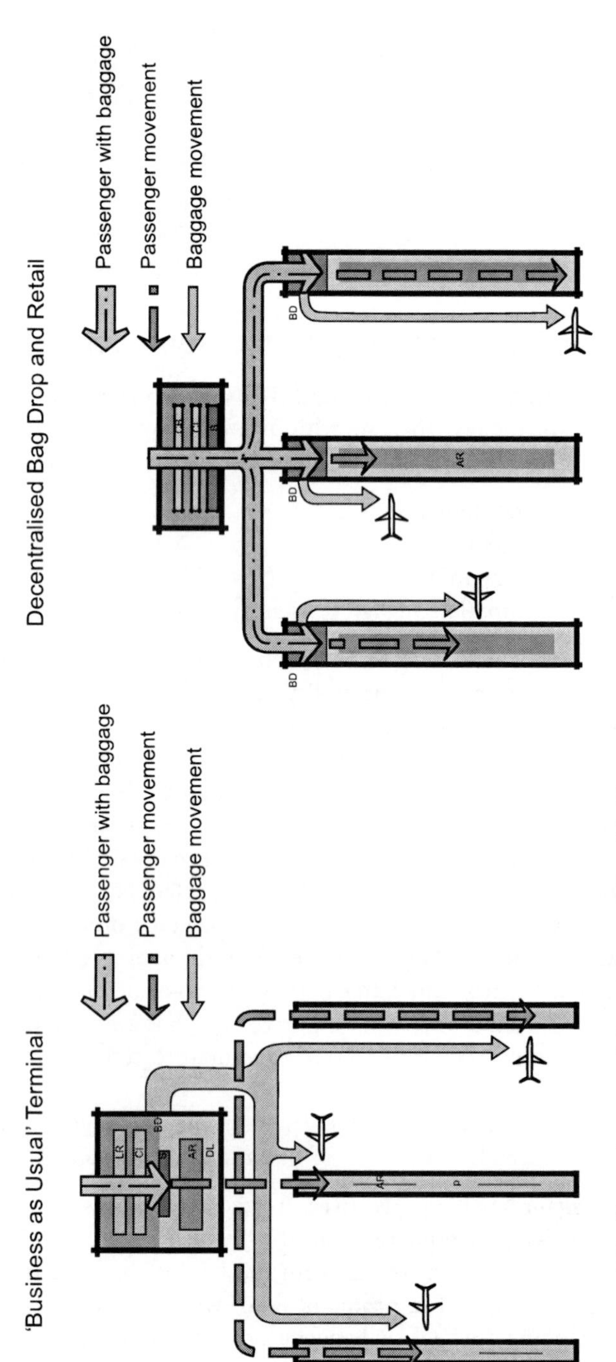

4.1 Current versus future low-cost terminal baggage process.

components are likely to have flight open times of, say, 3 hours but the baggage may be within the airport baggage system for, say, 7 hours or more. Again in this latter situation the correct quantity of costly resources is required for the longer periods. Clearly these ranges are extreme but are commonplace. New baggage systems need to reduce and optimise the baggage number of operatives required in an effort to reduce major operational (on-going as distinct from capital/initial cost) expenditure.

2. **Manual handling injuries.** Staff physical injuries occur mainly in two zones. The first is the check-in zone, where check-in operatives lift heavy baggage that is yet to be weighed and labelled on to the weigh conveyors. Here injuries are usually twist injuries to the lower back and can result in staff remaining off work for weeks at a time, which is costly to airlines and very bad for the long-term health of the staff. The most common injury zone is associated with staff working within the baggage hall and apron environment. As the baggage system requires baggage to be unloaded manually from the flight chutes, laterals and racetracks and then loaded into carts or universal loading devices (ULDs), this movement is naturally susceptible to injury. The three types of injury that occur are repetitive lifting strains, lifting of excessively heavy baggage which that not been labelled appropriately or twist and stretch injuries, or combinations thereof. All are serious, although the latter is the most commonplace and can lead to serious long-term problems for injured staff.

Next-generation baggage systems should try to incorporate new technology, which should address these issues head on. The problem in the past has been that the available technology to date only partly addresses the manual handling injury issue and has often been clumsy and difficult to use so that the operatives prefer not to use them despite being trained to do so.

Operationally proven new airport technology is available now that addresses manual handling injuries and has the real potential to reduce airport and airline operating expenditure. The technologies include robotic handling equipment and snake type feeder conveyors, which will load baggage into carts and ULDs. It should be noted that the use of robotics can work particularly well where early flight builds are required, e.g. long-haul traffic. Figure 4.2 shows a modern robotic installation. There are a number of technologies being investigated at the moment that look to improve operational efficiency and health and safety; these include the automatic movement of empty and full ULDs within the baggage hall environment and automatic movement of ULDs between the baggage hall and the aircraft. These ULD movement technologies are 'cutting edge' but are not

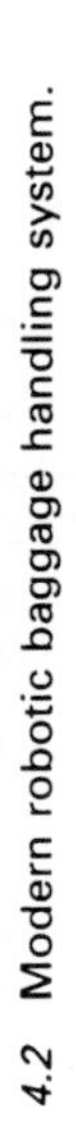

4.2 Modern robotic baggage handling system.

currently proven for use within the airport environment, but clearly show merit to develop further.

4.6 Advantages and disadvantages of the various hold baggage screening (HBS) processes and locations

As each airport differs in its design and traffic characteristics, the screening method applied should be a system that suits local conditions, although all should adhere to the requirements of ICAO Annex 17 and the national mandatory requirements in this regard. Each airport needs to consider the impact of cost, capacity and local operating conditions when developing appropriate solutions for both the location of screening and the methods/ technologies to be used.

In determining the most effective and efficient solution, the following principles should be applied:

- The system must provide screening solutions for both original departing and transfer baggage.
- The impact on valuable airport and terminal capacity should be minimised, while maintaining acceptable security and customer service standards.
- Investment in buildings, equipment and personnel should be minimised, while maintaining acceptable security and customer service standards.
- There should be minimal inconvenience to the airport operation and the travelling public both during construction and installation and day-to-day operation.

Locations of baggage screening systems may include:

- off-airport check-in (city centre, hotels, etc.)
- sterile terminal complexes
- sterile security area before check-in
- screening in front of check-in
- screening devices at or behind check-in
- screening downstream or 'in-line' within the baggage system.

4.6.1 Off-airport screening

As the title suggests, this type of screening is conducted away from the airport. The location can include railway stations, hotels and even major shopping complexes. It is common to have small automated baggage systems in railway stations supporting the screening process (if used). It is essential that screening away from the airport should be controlled very carefully using mechanical means and failsafe protocols to ensure that baggage cannot be tampered with, having been screened and then awaiting subsequent transferral to the airport. The quality of screening should be identical to the processes used within the airport and should incorporate screening equipment and screening protocols to the national standard.

Advantages are: (1) an improved service for downtown passengers; (2) permit segregation of higher risk flights and passengers only if screening is completed within the off-airport processing area; (3) space used off-airport is equal to space saved in the airport terminal; (4) airport users are not disrupted during the fit-out phase of off-airport facilities; (5) the potential to raise the profile screening precautionary measures.

Disadvantages are: (1) new areas needed for off-airport facilities; (2) secondary costs associated with new off-airport operation; (3) synchronised airport and off-airport operation required, both in terms of baggage movement and passenger movement; (4) baggage that has been screened must be held in secured storage between off-airport and airport terminal facilities; (5) transfer passengers cannot be processed through the off-airport facility; (6) utilisation of screening equipment is low because of fragmented operations.

4.6.2 Sterile terminal

The entire passenger terminal building is declared a sterile zone and all baggage, goods and all persons, passengers' staff and visitors entering the building must be screened to the same standard as passenger pre-board screening. This involves the creation of a sterile area at the boundary of the passenger terminal building, and can lead to prolonged queuing on the terminal forecourt (public access roads) areas.

Advantages are: (1) centralised screening maximises equipment and manpower utilisation; (2) no interference with existing check-in process or equipment; (3) passengers see security as high profile; (4) probably easier to incorporate new technology as it becomes available, as equipment is not linked into baggage systems.

Disadvantages are: (1) all items entering the terminal are screened, although the majority of them may not be related to checked baggage or a threat to aircraft; (2) complete sterility very difficult to achieve unless all goods and consignments are subject to security controls – all personnel must also be screened; (3) exits must be controlled to prevent unauthorised access; (4) large screening areas required at each entrance to the building, which may have to be constructed in passenger drop-off zones, which will need to be relocated; (5) disruption and capacity loss during construction; (6) possibility of passengers queuing three times (terminal entrance/check-in/ government inspection services); (7) high-profile passengers from ethnic groups or at-risk airlines are at an increased risk of terrorist attack during extended queuing at the entrance to terminal buildings; (8) passengers arrive either earlier or spend less time in commercial facilities; (9) in order to maintain terminal service standards, additional screening points may need to be provided, which will increase capital and operating costs; (10) can only be used for originating bags, not transferring baggage; (11) a suspect bag cannot be moved after screening, so requires terminal evacuation.

4.6.3 Sterile security area before check-in

This involves the creation of a sterile area either at the boundary of the check-in area or at several smaller zones within the check-in hall. Passengers and their carry-on bags should also be screened, or the checked bags wrapped or banded immediately after screening to prevent items being introduced after screening; alternatively the passenger and bag can be escorted to the check-in desk by airline or airport security personnel. Note that soft-sided zipper bags are difficult to secure adequately with banding machines.

Advantages are: (1) centralised screening provides better utilisation of equipment and manpower; (2) can be used for high-risk security flights; (3) other security procedures (profiling) can be carried out while passengers are queuing at the screening point; (4) passengers see security as high profile; (5) no interference with the existing check-in process or equipment; (6) no further hold baggage procedures for passengers required after check-in; (7) no interference with the existing baggage handling system; (8) probably easier to incorporate new technology as it becomes available as equipment is not linked into baggage systems.

Disadvantages are: (1) passenger and cabin bags must be screened simultaneously to prevent transfer of unscreened goods into checked baggage post-screening; (2) exit from the sterile zone must be controlled; (3) possibility of passengers queuing three times (entry, check-in, carry on); (4) may reduce attractiveness of commercial facilities to the non-traveller; (5) the large screening areas required at the entrance to check-in zones will reduce terminal capacity by up to 20%; (6) probable disruption and capacity loss during construction; (7) in order to maintain terminal service standards, additional screening posts may need to be provided, which could increase capital and operating costs; (8) when several check-in zones are in use, passenger queuing areas will need to be controlled to ensure efficient passenger flows to designated check-in points; (9) can only be used for originating bags not transfer; (10) a suspect bag cannot be moved after screening, so requires terminal evacuation.

4.6.4 Screening in front of check-in

The checkpoint is located directly in front of the airline check-in counters. All check-in baggage, luggage and other objects that possibly may not be permitted as hand baggage in the passenger cabin must be screened. If this approach is adopted and the baggage is to be returned to the passenger after screening for transport to the check-in counter, stringent measures must be taken to prevent passenger transferring unscreened items to bags that have been screened and to ensure that any unscreened bags are not, then, checked in as hold baggage.

Advantages are: (1) only checked baggage is screened; (2) can be used for flights with enhanced threat; (3) passengers see security as a high profile; (4) no impact on non-travelling public; (5) no additional passenger queuing required; (6) passenger and bag easily re-united if a hand search is required.

Disadvantages are: (1) careful surveillance required to avoid interference with screened baggage; (2) passenger screening process can be conducted during queue dwell time; (3) dedicated additional space required for screening equipment and process, including a dedicated hand search area for a minimum of 10% of checked bags; (4) projected loss of capacity up to 20% or a corresponding increase of pre-check-in space required; (5) cannot be used for transfer baggage; (6) a suspect bag cannot be moved after screening, so requires terminal evacuation.

4.6.5 Screening during check-in

Baggage is screened during or immediately after the check-in process. Screening equipment can be integrated into each individual check-in desk, on the feeder bag tag belt or in a security zone located at the rear of the check-in desks. These installations typically have conventional X-ray equipment installed, which requires a manual search of a minimum of 10% of bags. This search can take place either adjacent to the check-in or in a special screening area close to the check-in area.

Advantages are: (1) only checked baggage is screened; (2) passengers see security as high profile; (3) no effect on non-travelling public; (4) no additional passenger queuing required; (5) although passenger processing times at check-in may increase, this process may not involve a loss of capacity at some airports.

Disadvantages are: (1) major capital costs – each check-in desk needs screening equipment to be installed; (2) possible requirement for new check-in desks if existing desks cannot be retrofitted; (3) will require modification to the baggage handling equipment at the check-in desk; (4) may require changes to the check-in process to deal with baggage first so it may be screened while other passenger check-in processing is completed; (5) possible reduction of number of check-in desks per island or terminal (linear arrangement); (6) check-in transaction times could be increased; (7) operators under pressure to screen bags quickly; (8) cannot be used for transfer baggage; (9) additional space required for manual search adjacent to or behind desks; (10) a suspect bag cannot be moved after screening, so requires terminal evacuation.

4.6.6 Manual screening

Screening is carried out at dedicated locations. This can be either at an off-airport location or co-located by the departure check-in area. Manual searching is a resource-intensive task and the most appropriate use is for operations with low volumes. It requires a significant number of well-trained and motivated staff, often fully employed only for short periods of time and dedicated areas set aside for the search process. It can be conducted in a mobile (dedicated screening vehicle) facility and used as the final arbitration for other techniques.

Advantages are: (1) centralised locations may require less space for manual screening compared with a search location behind each desk; (2) centralised locations may require fewer personnel to be deployed for manual screening duties compared with a search location behind each desk; (3) direct contact with baggage, and considered very effective and reliable for most articles; (4) bag screening takes place during the check-in transaction and the passenger can be relatively easily available if manual screening is required; (5) screening can be conducted at the aircraft side in a dedicated screening vehicle; (6) it is the final arbitration for all other techniques.

Disadvantages are: (1) baggage needs to be taken to the search area; (2) not as easy for search teams to communicate with the check-in staff when they are in separate locations; (3) requires training on concealment techniques; (4) total reliance on human factor issues – search must be thorough and efficient; (5) space requirements for manual screening may dictate the need for new construction to replace lost capacity; (6) not fully effective for complex articles with electronic components; (7) only practical for small volumes/throughput; (8) only possible with passenger present; (9) labour resource-intensive operation; (10) a suspect bag cannot be moved after screening, so requires terminal evacuation.

4.6.7 Screening downstream 'in-line' within the baggage system

Screening of checked baggage is carried out in the baggage sorting area or within the baggage handling system. These systems typically use explosives detection system (EDS) X-ray equipment.

Advantages are: (1) current check-in procedures not affected; (2) no extension of public areas of building required; (3) only checked hold baggage is screened; (4) operators are under less pressure to screen bags quickly; (5) no effect on non-travelling public and commercial revenue; (6) baggage is security controlled after check-in; (7) a 'suspect' bag having already been handled can be moved if necessary, so preventing terminal evacuation; (8) centralised screening within the baggage system will maximise machine utilisation; (9) can be used for transfer and originating bags.

Disadvantages are: (1) passengers are not aware of security measures; (2) could require extensions to baggage sort areas to accommodate equipment/screening rooms; (3) problems could occur when reuniting passengers and their baggage for manual screening if required; (4) difficulties with reuniting passengers and their bags may cause delays to flights, especially when near to departure time; (5) may require significant changes to the baggage sorting system with cost/capacity implications; (6) screening equipment may require slower baggage belt speeds, which may reduce baggage system capacity.

There are three types of IATA recommended hold baggage screening processes for in-line security that should be adopted, see Figs 4.3, 4.4 and 4.5. The type of security process should align with the size of the airport and the volume of baggage processed per year.

Important note. It should be noted that national legislation should be observed, which may dictate that specific screening machines are used and also the route they take. If in doubt see advice from the national governmental security advisor. As an example, FAA/TSA compliant airlines will be required to process baggage through specific certified types of X-ray screening equipment.

4.6.8 Certified explosives detection system (EDS) lobby installations

This is certified EDS equipment meeting the US TSA Explosive Detection System criteria, located in the check-in lobby area, operated either as a 'Drop and Go' screening point or for passengers referred from check-in. Currently this is limited to computed tomography (CT) equipment designed originally for integration with baggage handling systems. CT systems can be operated in automatic mode but have a moderate to low throughput when

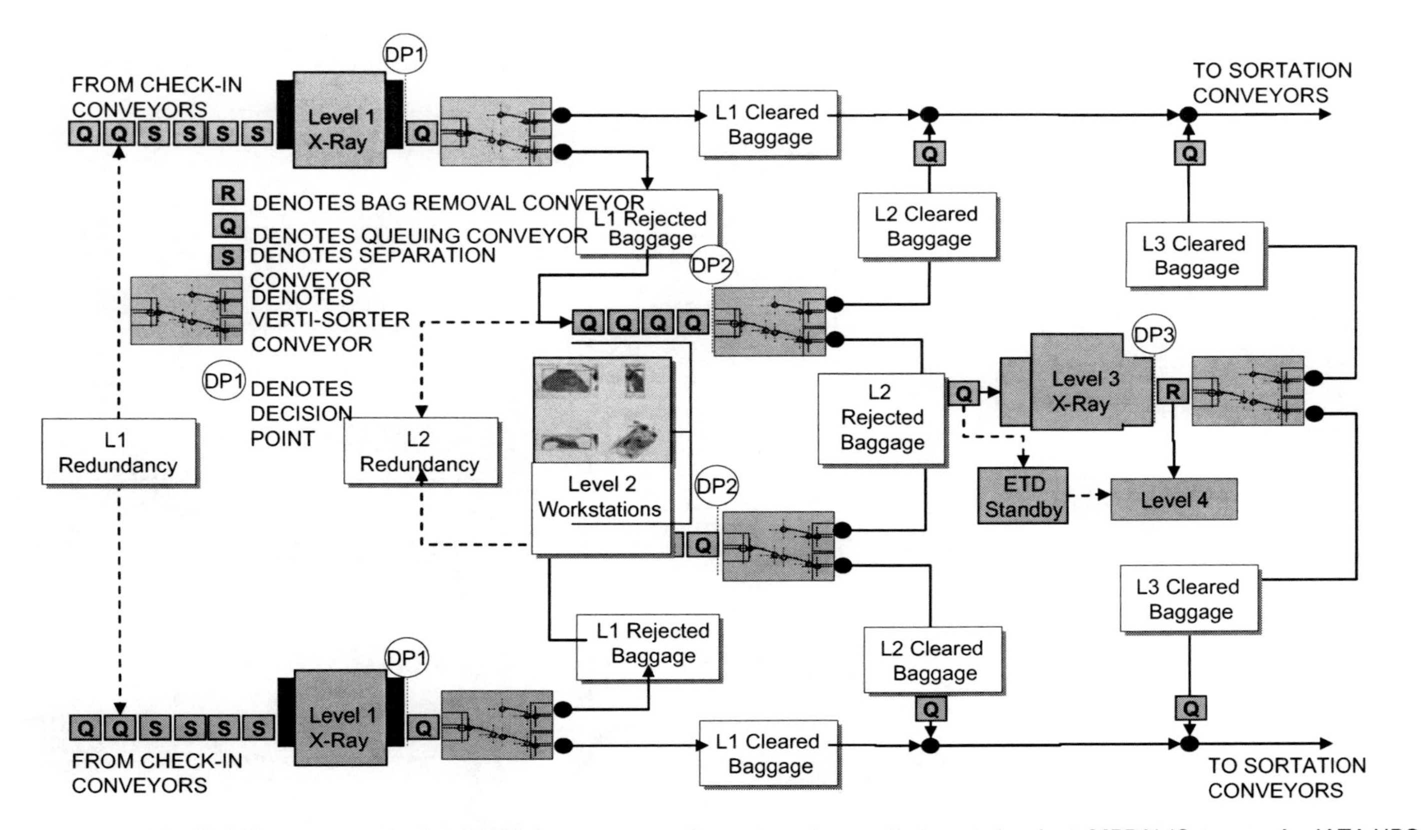

4.3 Hold baggage screening (HBS) departures configurations for small airports (under 1 MPPA) (Category A – IATA HBS Processes; courtesy of Norman Shanks).

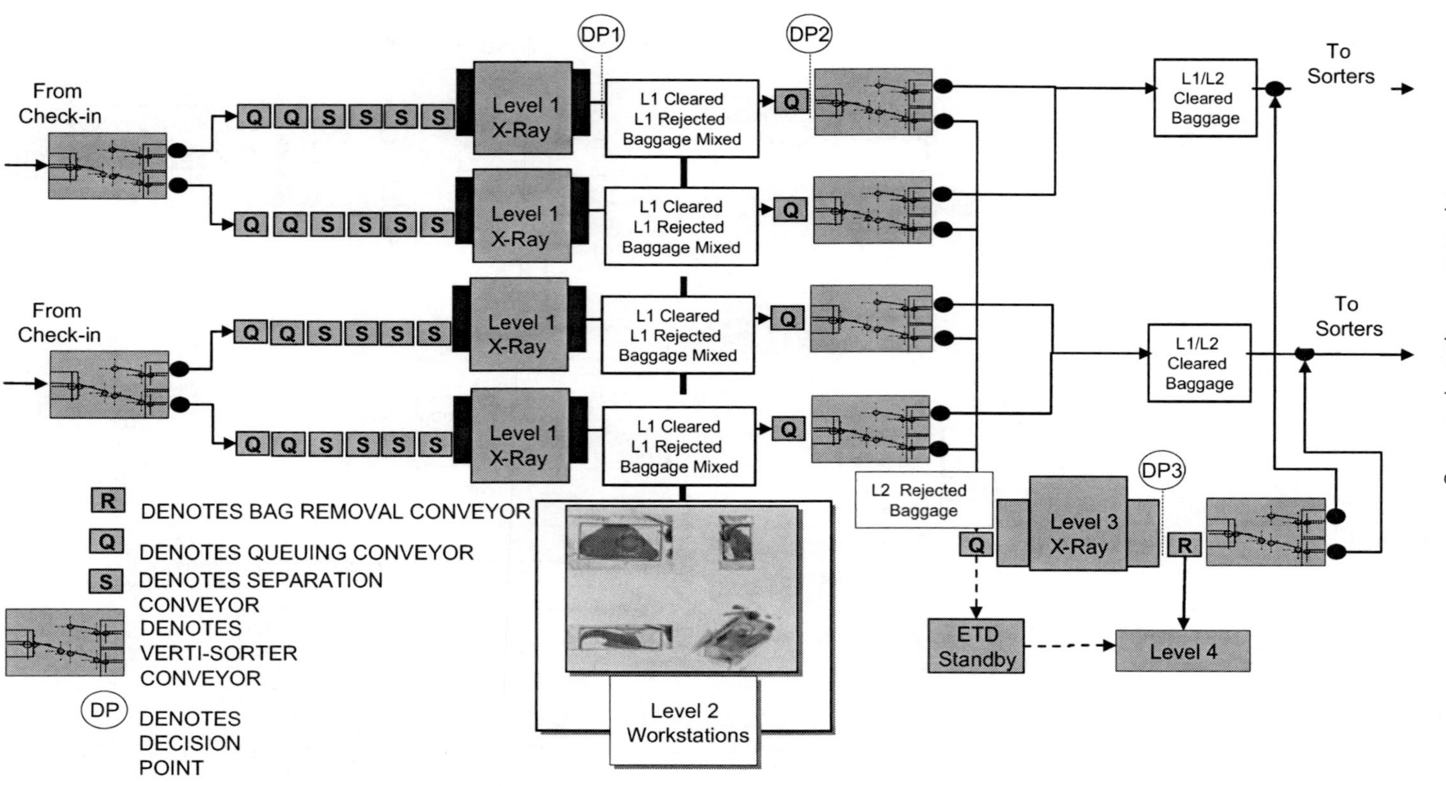

4.4 Hold baggage screening (HBS) departures configurations for medium airports (over 1 MPPA and under 5 MPPA) (Category B – IATA HBS Processes; courtesy of Norman Shanks).

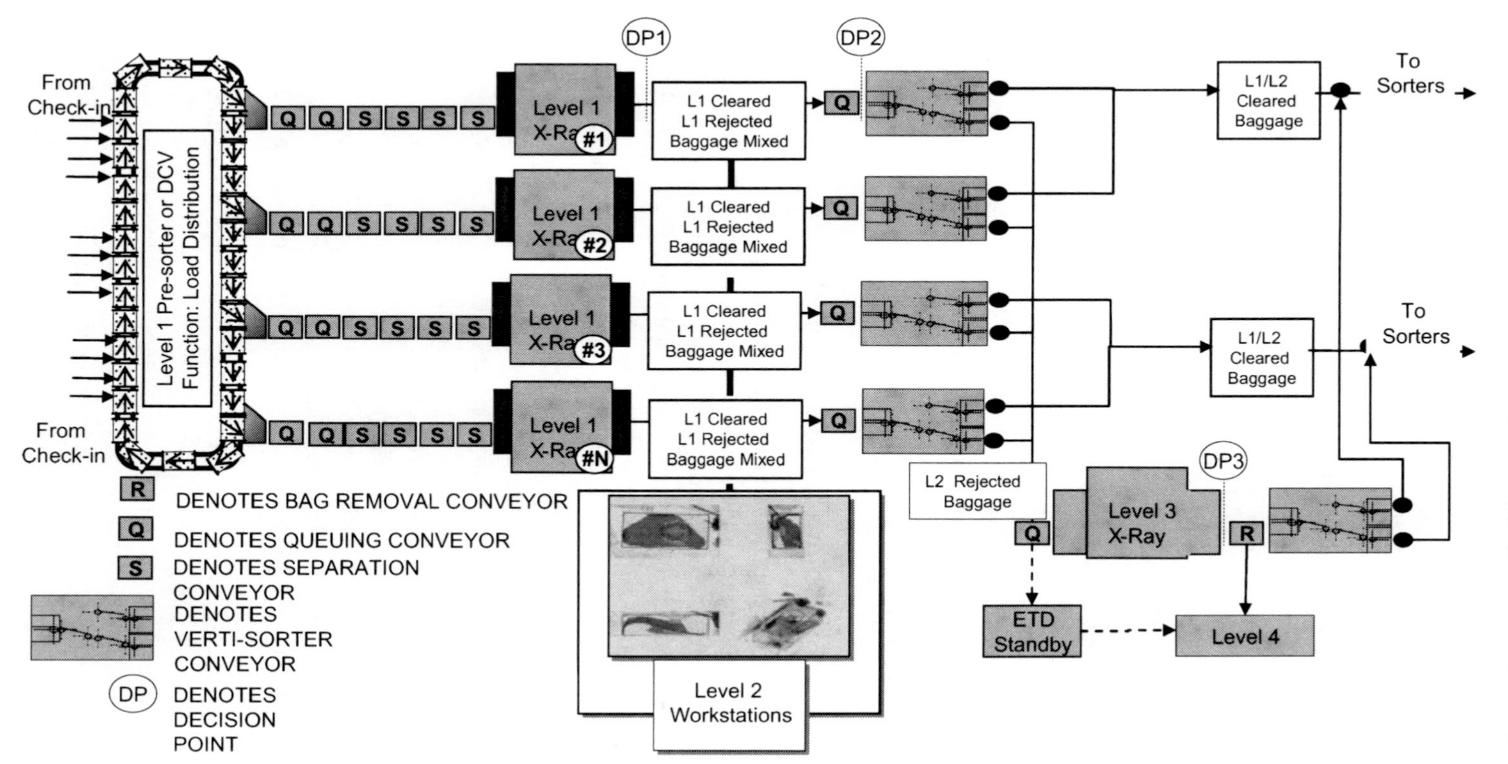

4.5 Hold baggage screening (HBS) departures configurations for large airports (over 5 MPPA) (Category C – IATA HBS Processes; courtesy of Norman Shanks).

used in stand-alone configurations and are not practical as single solutions for 100% screening at moderate to large airports in this configuration. Has a moderate false alarm rate (25% +), which requires human image analysis of complex CT images of alarmed bags.

Advantages are: (1) equipment meets the US federal requirements; (2) only checked baggage is screened; (3) can be used in automatic mode enabling operators to concentrate on rejected bags; (4) passengers available if required to be reunited with their bags.

Disadvantages are: (1) high cost of equipment (not relevant to US operations – government funded); (2) moderate to low throughput; (3) multiple detection configurations can lead to different detection standards; (4) moderate to high false positive alarm rates; (5) large numbers of high capital cost equipment required for moderate to large airports; (6) additional lobby or check-in space required or corresponding loss of capacity – not practical for moderate to high airport 100% screening; (7) high reliance on human factors – operator's skills in interpreting complex CT images; (8) a suspect bag cannot be moved after screening, so requires terminal evacuation; (9) cannot be used for transfer bags.

4.7 Manual handling baggage hall design

The design of the baggage hall will need to be carefully planned. It is a fact that the location, plan area and volume and connectivity of the baggage hall will need to be fully considered, even at initial masterplan stages and subsequently refined at each major design gateway. Figure 4.6 shows a number of options for siting baggage hall systems.

The baggage hall design will need to consider the following attributes when designing a baggage system:

- Confirm if departures flows are to be processed.
- Confirm if transfer flows are to be processed.
- Confirm if arrivals flows are to be processed.
- Confirm flow integration requirements.
- Confirm check-in to baggage hall processing times.
- Confirm check-in to aircraft connection times.
- Confirm short-term and long-term masterplan alignment.
- Confirm system maintenance strategy (important for volume and access requirements).
- Confirm the manual handling/automatic (e.g. robotic handling strategy).
- Confirm the airline and handler space and user interfaces.
- Confirm hold baggage screening protocols.
- Confirm out-of-gauge (OOG) processing requirements.

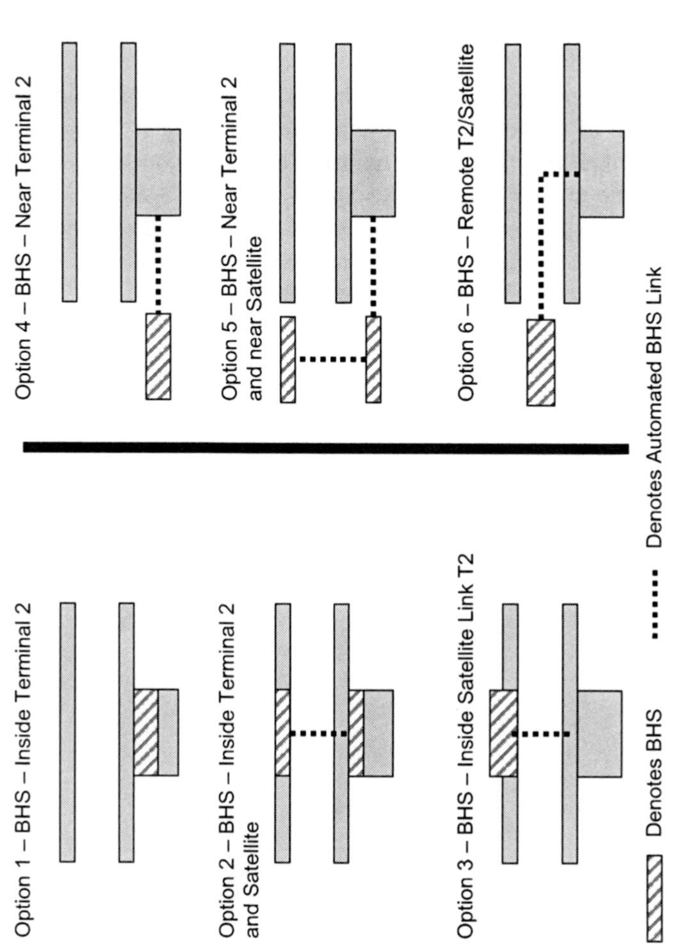

4.6 Master plan baggage handling system (BHS) location options.

- Connectivity requirements of the baggage systems between terminals and piers and satellites and in some instances the head of stand.
- Apron road layout interfaces.
- Baggage hall passing and parking lane and manoeuvring requirements.
- Tug charging facilities.
- Universal loading device (ULD) sizes, types and opening requirements.
- Tug sizes and power type.

Where manual handling is permitted the airport designer should consider the implications of the location of the manual handling interfaces and the loading and unloading levels and baggage handling conveyor equipment speeds. It is the author's recommendation that professional health and safety advice is sought to define manual interface loading levels and to define the suitable manual handling lifting aid and manoeuvring devices that could be selected that best fit the proposed operation.

5

Airport apron, runway and taxiway design

Abstract: This chapter reviews airport apron, runway and taxiway design. After outlining function areas of the apron, it discusses aircraft stands and multiaircraft ramping systems (MARS) as well as passenger airbridges. It reviews key design issues such as levels of passenger service and the different requirements of low-cost carrier terminals versus full-service legacy terminals. The chapter concludes by discussing runway components.

Key words: airport apron, runway, taxiway, multiaircraft ramping systems (MARS), passenger airbridges.

5.1 Function areas of the apron

The capacity of existing and future airport facilities must be estimated in order to establish the current capability to accommodate forecasted traffic growth. A simple breakdown of airport capacity is as follows:

- runway capacity, in terms of the aircraft movement rate in a given period
- apron capacity, in terms of the number of aircraft stands available
- terminal capacity, in terms of passenger and baggage throughput per hour or cargo throughput per hour
- landside access capacity, in terms of the number of passenger/vehicle throughput per hour.

Runway capacity is a product of many features of the airport and its surroundings, namely length and width of the runway, capacity of the airspace, the aircraft codes that it can service, take-off and approach terrain, general prominent weather conditions, coding of the airport infrastructure such as fire safety equipment provisions and landing guidance systems such as instrument landing systems (ILS) and noise restrictions.

Apron capacity is defined by the number of simultaneous aircraft stand centrelines that can accommodate the busy hour periods of departing or

Table 5.1 Sample from ICAO Airport Reference Code showing letters A to E for wing span and wheel span

Code letter	Wing span	Wheel span
A	Up to but not including 15 m	Up to but not including 4.5 m
B	15 m up to but not including 24 m	4.5 m up to but not including 6 m
C	24 m up to but not including 36 m	6 m up to but not including 9 m
D	36 m up to but not including 52 m	9 m up to but not including 14 m
E	52 m up to but not including 65 m	9 m up to but not including 14 m
F	80 m box (specifically A380)	

arriving passenger flows. A sample from the ICAO Airport Reference Code is denoted in Table 5.1. This table shows the code letter for wing span and wheel span codes A–E inclusive. In practice, wing span is the dominant criterion that will define apron capacity for a given area, along with the mandatory interstand road clearances and wing tip clearances, all of which dictate the length of the contact piers and satellites.

5.2 Aircraft stands and multiaircraft ramping systems (MARS)

Figure 5.1 shows the arrangement of a pair of typical push-back parallel stands. The stand shown on the left (item H) is a typical single centreline push-back stand whereas the stand on the right (item F) is a multiaircraft ramping system (MARS) stand arrangement, not shown to scale. The MARS stand allows either one Code E/F aircraft to be accommodated on the main central centreline or to accommodate up to two Code C aircraft simultaneously on the left and the right stand centrelines within the confines of the stand safety lines (item I). This arrangement can be very efficient and any airport planner should review the pros and cons of this before committing to layout.

It should be noted that international recommendations or best practice exists with respect to the placement of the fuel hydrant positions. Fundamentally they should not be located beneath the engines of aircraft when the aircraft are in the correct parked positions. This sounds straightforward but actually can be very difficult to achieve with a wide range of aircraft. The airport planner may decide it is more cost effective and flexible to use the lower CAPEX solution of mobile fuel trucks. The main additional advantage with this option is that the apron can be changed in configuration very easily. The small Category A type airports tend to use fuel trucks for this reason, as change or growth is almost inevitable and the volume of traffic does not warrant the major capital expenditure necessary to achieve in-ground fuelling system infrastructure.

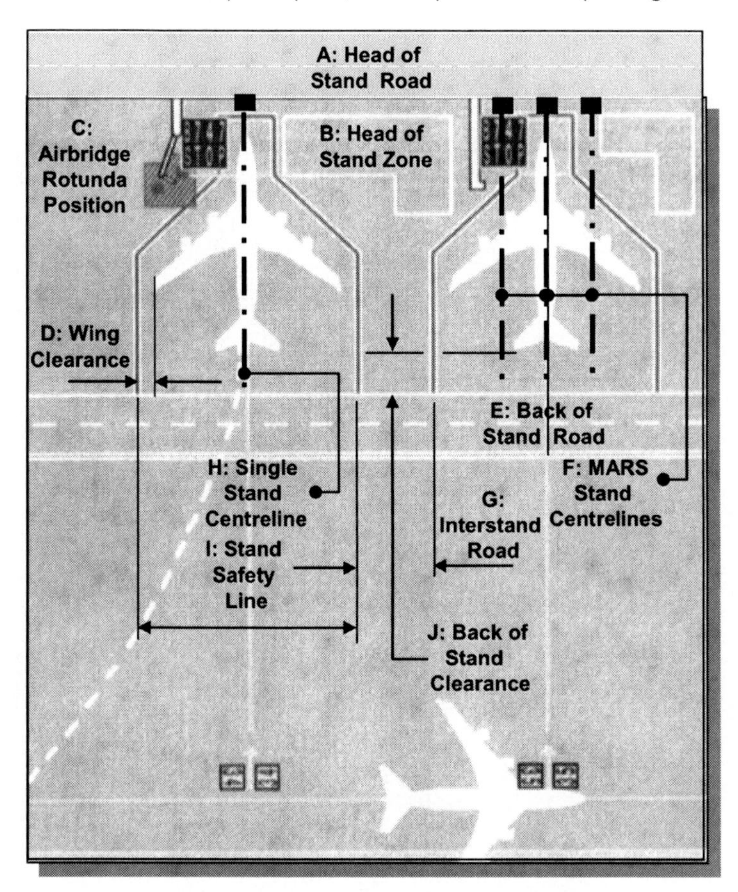

5.1a Aircraft stands: single centreline and MARS centrelines.

With reference to Fig. 5.1(a) and (b), the aircraft stand comprises many different zones. These include:

Zone A: head of stand road. This is a road predominantly used by aircraft servicing vehicles to gain access to the aircraft along the length of the pier or satellite. The road can be used by aircraft push-back tugs as well. There are occasions where it is not necessary/possible to provide a head of stand road. When this occurs it is normal to provide a back of stand road as a second best choice.

Zone B: head of stand zone. This area is used to accommodate a wide range of equipment from push-back tugs along the stand centrelines to the location of the fixed ground power positions and even dolly parking. The zone also permits the installation of aircraft parking positioning equipment such as parallel axis parking aids (PAPA) and AGNIS parking equipment or the more modern radar-based systems such as SAFEDOCK. It should be noted that the parking of dollies alone in

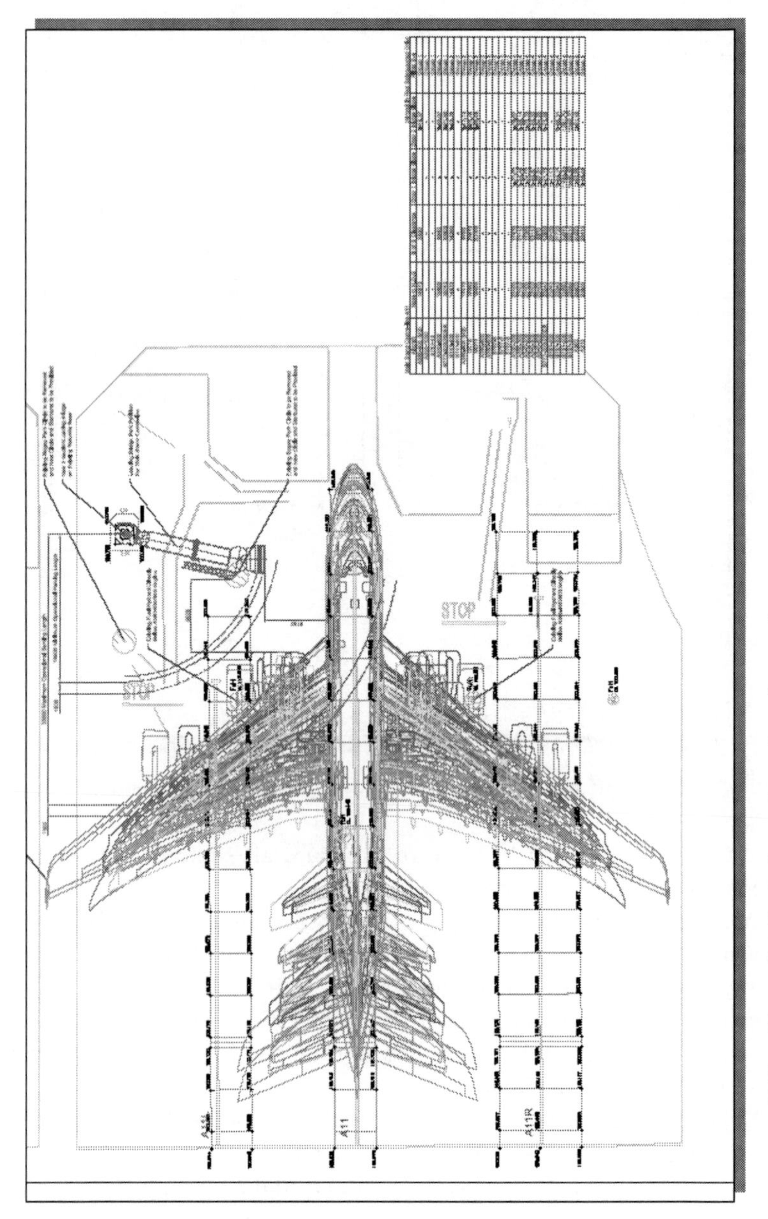

5.1b More detailed view of an aircraft stand.

designated safe areas away from jet intakes at the head of the stand area may be permitted, but dollies and ULDs or wessex type trucks should not be parked in this zone. Loose ULDs or loose bags within wessex type loose load baggage carts can be sucked into the path of aircraft engines.

Zone C: airbridge rotunda position. This is the fixed pivotal point of an airbridge. It is the point by which airbridges rotate and elevate about their normal axis of movement (2 degrees of freedom excluding the telescopic section movement). This position needs to be carefully calculated to ensure that the optimum length of airbridge is provided and that all aircraft can be served with a reasonable length/priced airbridge. The position will contribute to the depth of the stand and the length of the pier. It is for this reason that they should be carefully positioned to ensure that aprons are not too costly to provide.

Zone D: wing clearance. This is the distance from the widest aircraft wing tip (negatively mis-parked off centreline by 600 mm) to the edge of the stand safety line (**Zone I**). According to ICAO Annex 14 Volume 1 Clause 3.13.6, Recommendation –

An aircraft stand should provide the following minimum clearances between an aircraft using the stand and any adjacent building, aircraft on another stand and other objects:

CODE Letter	*Clearance*
A	*3 m*
B	*3 m*
C	*4.5 m*
D	*7.5 m*
E	*7.5 m*
F	*7.5 m*

When special circumstances so warrant, these clearances may be reduced at a nose in aircraft stand, where the code letter is D, E or F:

a) Between the terminal, including any fixed passenger bridge, and the nose of an aircraft; and

b) Over any portion of the stand provided with azimuth guidance by a visual docking guidance system.

Note – On aprons, consideration also has to be given to the provision of service roads and to the provision of manoeuvring and storage area for ground equipment (see the aerodrome design manual, Part 2, for guidance on storage of ground equipment).

Zone E: back of stand clearance. This road has the same function as the head of stand road defined in Zone A above. This road is often not provided. The back of stand road is inherently less safe to operate than the head of stand road and it is for the reason that they are the second choice of access to the stands. The back of stand road requires very effective ground communications between tower, aircraft and ground vehicle operations to enable them to operate safely, particularly on busy airfields. Refer to ICAO Annex 14 Volume 1 for more details.

Zone F: MARS stand centrelines. These stands will usually accommodate multiple stand centrelines within the confines of the stand safety line or stand perimeter, usually three stand centreline markings are provided, though two or even four stand centrelines can be provided. It should be noted that four stand centrelines are less common and the fourth centreline is usually a diagonal centreline, which typically would be at 35 degrees to the main centrelines. This configuration can sometimes allow for the installation of a Code C and a Code E to be served on the same stand perimeter.

Zone G: interstand road. This road allows the aircraft servicing vehicles including passenger movement vehicles to gain access to the inner stand areas. It is commonplace/essential for interstand roads to be located on both sides of the aircraft stand as aircraft require both the starboard and port sides to be serviced.

Zone H: single stand centrelines. These are stands that have a single centreline used to service aircrafts from Code A up to Code F inclusive. It is commonplace for single stand centrelines to be used for aircraft up to Code D. Beyond Code D the use of MARS stands should be considered.

5.3 Passenger airbridges

Passenger airbridges are used for a number of reasons, some of which are listed below:

- enhanced passenger service (smooth transition from aircraft to airport building levels/good controlled walking environment)
- enhanced passenger safety (no walk across aprons/head of stand roads/ change of building levels/safe walking condition in poor weather)
- speed of embarkation and disembarkation (which is argued by short-haul low-cost carriers)
- speed of aircraft turnaround (which can be very significant for long-haul operations).

Since the late 1990s the influx of lower-cost airlines in some countries has steered airports away from providing airbridges, as many of these airlines, although not all, will tend not to use them because they require a payment to the airport operator for their use. Instead, many of these airlines prefer to use their own passenger stairs built into the aircraft on both the forward and aft doors.

It is very prudent to design all stands with the capability to accommodate aircraft passenger airbridges. As the fleets of these lower-cost airlines grow over time they will upgrade to Code D and E plus aircraft. This will mean aircraft turnaround time/critical path is dictated by the ability to embark and disembark passengers from the aircraft. This in practice is likely to mean the installation of aircraft passenger load bridges. Airports processing only small Code C type aircraft (e.g. B737 and below) arguably do not need passenger airbridges to obtain a fast turn-around time, but may decide that they are required to obtain a good passenger service level.

The following general rules of thumb apply in the decision to install airbridges.

- All stands should be designed to accommodate airbridges.
- Airports processing predominantly legacy airlines will need to provide airbridges to obtain the required passenger service levels.
- Airports processing predominantly low-cost carriers should first contractually understand if the airlines will pay for the use of the airbridges.
- Airports processing less than 5 MPPa should at least safeguard the use of airbridges.
- Airports that often operate in adverse weather conditions should consider operating airbridges for passenger safety reasons.
- Airports operating greater than 5 MPPa throughput of Code D or higher aircraft should install passenger airbridges.
- Airports operating Code F aircraft should consider the installation of two airbridges to service each Code F aircraft.

5.4 Levels of passenger service

The use of airbridges raises the question of overall levels of passenger service that such facilities should meet, particularly in ensuring the smooth flow of passenger traffic and appropriate levels of comfort. Historically IATA has defined the following scale, which allows comparison of a range of level of service measures categorised from best 'A' through to worst 'F':

- Service level 'A' should provide an excellent level of service, where passengers will flow freely throughout the terminal complex and where there are excellent levels of comfort for passengers.
- Service level 'B' should provide a high level of service where the passengers should experience stable passenger flows with few delays and a high degree of passenger comfort.
- Service level 'C' should provide a good level of service. Again the passengers should witness stable flow conditions, but where delays could occur these delays are deemed to be reasonably acceptable. The

Table 5.2 Recommended levels of service in different terminal areas

	Bradley level of service standards (m^2/pax)					
	A	B	C	D	E	F
Check-in queue area	1.8	1.6	1.4	1.2	1.0	
Wait/circulate area	2.7	2.3	1.9	1.5	1.0	
Hold room	1.4	1.2	1.0	0.8	0.6	System
Bag claim area (does not include reclaim unit area)	2.0	1.8	1.6	1.4	1.2	failure

passengers should experience good levels of comfort within the passenger interfacing areas.

- Service level 'D' should provide an adequate level of service. There will be instances where passenger flow is seemingly unstable. Passengers with this service level will experience acceptable delays for short periods of time and passenger comfort levels will be adequate.
- Service level 'E' would provide an inadequate level of service to passengers where passenger flow is essentially unstable. Passengers will frequently experience unacceptable delays and inadequate levels of passenger comfort.
- Service level 'F' will provide unacceptable levels of passenger service. Passengers will frequently experience conditions of cross-flow and system/process failures will frequently occur and there will be frequent unacceptable passenger processing delays. Passengers would experience unacceptable levels of comfort.

Important note. It should be noted that irrespective of the service level selected, the designer/architect has the responsibility to ensure passenger flows are safe both in normal operating conditions and during periods of evacuation. Passenger movement simulation tools and environmental/fire simulation tools should be used to confirm safe operation of the terminal during all likely operating periods and scenarios.

Service level 'C' has been historically recommended as the minimum design objective as it denotes good service at a reasonable cost. Service level 'A' is seen as having no upper set of limits. With respect to terminal building, examples of level of service metrics are denoted in Table 5.2 for use when planning terminal and pier/satellite building infrastructure. The reader should also refer to the 9th edition of the IATA *Airport Development Reference Manual* Chapter 9, and in particular table F9.2, for further information.

Table 5.3 Examples of low-cost airline developed/used terminals

Airport	~% LCC traffic 2009
Stansted	85
Amsterdam Schiphol Airport	6
Billund – Denmark	5
Cologne	6
Dublin	36
Geneva	16
Frankfurt Hahn	99
Las Vegas McCarran	58
Luton Airport	86
Milan – Malpensa	4
Ottawa Airport	18
Palma de Mallorca	36
Phoenix Arizona (Sky Harbor)	80
Singapore (Changi)	0.4

5.5 Low-cost carrier terminals versus full-service legacy terminals

There are many factors that will dictate the characteristics of a terminal planned to be used by low-cost carriers and those used to service full-service legacy carriers. Examples of locations that process, or were developed to process, predominantly low-cost service airlines are denoted in Table 5.3.

Following study of a number of these European, Asian and American airports processing predominantly low-cost carrier airlines, the following observed airport architecture and system characteristics are evident (see Fig. 5.2):

- older infrastructure often reused where possible
- basic architecture (reuse of cargo buildings/simple steelwork/basic cladding systems)
- lower technology at check-in and within baggage hall (automatic baggage sortation but manual sortation often used)
- less IT infrastructure (does not include security – same as full service)
- lowest possible operating cost mandated while maintaining an effective operation
- longer walking distances often evident
- passengers often walk on apron to connect with aircraft
- no business lounges
- gate seating often limited
- call to gate messaging often controlled by airline directly, thereby negatively affecting retail sales.

Examples of locations that were developed for predominantly full-service

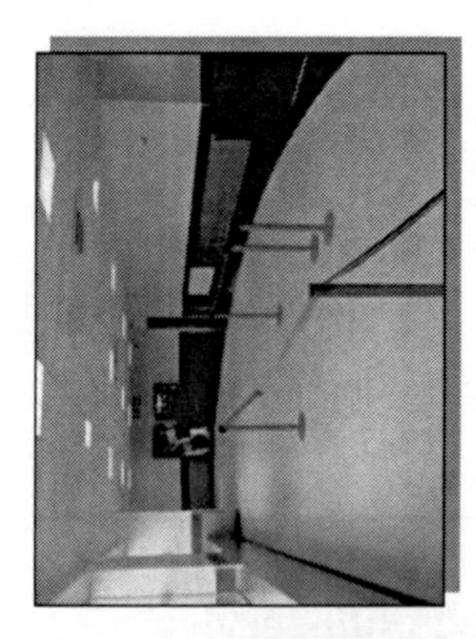

Gate room – no seats, minimal legal space

Passenger required to walk on apron to connect with aircraft

Terminal concourse area – no windows, converted cargo building

5.2 Examples of typical low-cost operator terminals.

Table 5.4 Examples of legacy airline developed terminals

Heathrow Terminal 5	Shanghai
Seoul	Hong Kong
Tokyo Narita	Taipei
Osaka	Manila
Nagoya	Bangkok
Beijing	Kuala Lumpur
Singapore	

airlines are denoted in Table 5.4. Airports predominantly processing full-service legacy airline operations tend to have the following characteristics (see Fig. 5.3):

- compliance with Level C or D IATA service level standard achieved or better
- often signature architecture evident, although not an essential component
- use of medium- to high-end automation for baggage processing
- transfer baggage processed
- early baggage storage where needed
- multiclass check-in and baggage sortation
- world-class passenger experience – e.g. fast, convenient check-in/airbridge to aircraft
- world-class retail and revenue
- low to medium airport operating cost.

Extensive study has shown that single-level buildings are generally far more flexible and can much more easily adapt to a change in traffic processing needs when compared with multilevel buildings. A good example of this is the Stansted airport terminal, which was originally planned to accommodate legacy long-haul operations when it opened. When the air traffic did not initially materialise the open-plan building concept was able to adapt easily when the short-haul low-cost traffic eventually took up the airport capacity infrastructure.

5.6 Runway components

The full categorisation of runways is given in ICAO Annex 14 Chapter 3.1, Runways. There are various factors that affect the design of runways or the number and orientation of runways within an airfield. The most important feature of a runway must be that it is usable and provides the necessary characteristics to permit the aircraft to take off and land efficiently and safely. A major factor in the orientation of runways is the direction of the

Bangkok

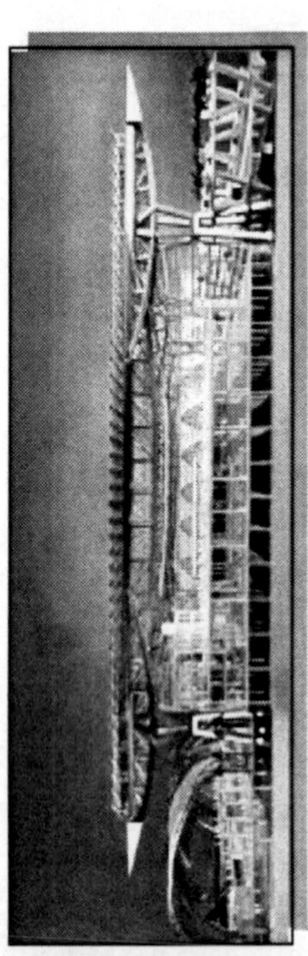

5.3 Examples of typical full-service legacy-cost terminals.

prevailing winds. Runways should ideally be orientated to run in line with the prevailing wind direction. This condition will reduce the difficulties associated with crosswind take-off and landing of aircraft. The length of the runway will be dictated by many factors, such as aircraft to be served from the airport, where some larger aircraft require longer take-off lengths and the general elevation/static pressure range of the airfield. The higher the altitude of the airfield the less dense is the air. This is a major factor to consider when designing the runway length. Each aircraft type to be served will need to be evaluated to ensure that the runway is of the correct length to allow safe positing and take-off length.

The core components of a runway are:

- Runway shoulder. These run parallel to the main runway surfaces and are present where the runway is of code letter D or higher. Where the runway is of code letter D or code letter E the combined width of shoulders and runway should not be less than 60 m. Where the runway is of code letter F the combined width of shoulders and runway should not be less than 75 m.
- Runway strips. This can be the runway and any associated stopways in combination.
- Runway turn pad. Where the end of the runway is not served by a taxiway or a taxiway turnaround and where the code letter is D, E or F, a runway turn pad shall be provided to facilitate a 180 degree turn of aircraft.
- Runway end safety areas. A runway end safety area shall be provided at each end of a runway strip where the code number for the runway is 3 or 4 or the code number for the runway is 1 or 2 and the runway is an instrument runway type.
- Rapid exit taxiway (RET). The RET is used by arriving aircraft. It allows aircraft of specific codes and specific set safe higher speeds to turn rapidly off the main runway having landed. The benefits are:
  (a) The aircraft fuel burn on the ground can be considerably less in some instances.
  (b) The main runway can accommodate more aircraft movements as it is not occupied by slow-moving aircraft.
  (c) The aircraft can achieve faster turn-around and thus efficiency can be high.
- Rapid access taxiway (RAT). The RAT is used by departing aircraft. It allows aircraft of specific codes and specific set safe higher speeds to turn rapidly on to the main runway from a taxiway. The benefits are similar to those witnessed with RETs.

The use of multiple runways is commonplace and possible where some airports dedicate operations to one runway for take-offs only and the other

Table 5.5 Width of runways according to ICAO definition

Code number	Code letter					
	A	B	C	D	E	F
1*	18 m	18 m	23 m	—	—	—
2*	23 m	23 m	30 m	—	—	—
3	30 m	30 m	30 m	45 m	—	—
4	—	—	45 m	45 m	45 m	60 m

*The width of precision approach runway should be not less than 30 m where the code number is 1 or 2.

runway for landings only. The distance between runways is dictated by the possibility of the presence of turbulence for neighbouring aircraft. ICAO Annex 14 spells out the recommended distances associated with various airfield code letter categorisations.

Where parallel non-instrument runways are intended for simultaneous use, the minimum distance between centrelines should be:

Distance between runway centrelines	Higher ICAO code letter
210 m	3 or 4
150 m	2
120 m	1

Where the simultaneous use of instrument rated runways is required the distances between runways is defined to be:

- 1035 m for independent parallel approaches
- 915 m for dependent parallel approaches
- 760 m for independent parallel departures
- 760 m for segregated parallel operations

In some instances these distances can be relaxed. The reader should revert to ICAO Annex 14 Clause 3.1.11 for definitions of situations where this may be possible.

The width of runways is defined within the ICAO table (here presented as Table 5.5).

The permissible and recommended parameters for all aircraft taxiways are defined extensively within ICAO Annex 14 Clause 3.8. The aircraft taxiway is effectively the service lane on an apron for aircraft. Aircraft use the taxiway to connect from the stand areas to the runway and vice versa. The width, length, clearances between apron features and the bend radii of taxiways are all categorised by ICAO in Annex 14.

Table 5.6 Aerodrome reference codes (ICAO table 1-1; reproduced with kind permission by ICAO)

Code element 1		Code element 2		
Code number (1)	Aeroplane reference field length (2)	Code letter (3)	Wing span (4)	Outer main gear wheel span* (5)
1	Less than 800 m	A	Up to but not including 15 m	Up to but not including 4.5 m
2	800 m up to but not including 1200 m	B	15 m up to but not including 24 m	4.5 m up to but not including 6 m
3	1200 m up to but not including 1800 m	C	24 m up to but not including 36 m	6 m up to but not including 9 m
4	1800 m and over	D	36 m up to but not including 52 m	9 m up to but not including 14 m
		E	52 m up to but not including 65 m	9 m up to but not including 14 m
		F	65 m up to but not including 80 m	14 m up to but not including 16 m

* Distance between the outside edges of the main gear wheels. Note – Guidance on planning for aeroplanes with wing spans greater than 80 m is given in the Aerodrome Design Reference Manual, Parts 1 and 2.

5.7 Restricted surface runways and taxiway ICAO Annex data

Chapter 2 of this publication explains the sequential steps that should be followed when developing runways and taxiways. The reader should refer particularly to step 3 'Understand the runway configurations options', step 4 'Set runway alignment' and step 6 'Taxiway planning when developing the runway and taxiway complex'. Airport planners usually have a preference to set the airport design constraints that are necessary working outwards, starting from within the centre of the airport complex (terminal buildings) and then towards the fringes of the airport boundary, constantly giving appropriate consideration to the runway and apron and ATC needs and limitations that may be evident. The airport planners will then at a high level collate all of the principal options developed to a common level of detail, and analyse these options to ensure that the all-round best solution can be selected using the techniques described in Chapter 2, Section 2.2, Masterplan evaluation techniques (evaluation criteria/pairwise/weighting).

The reader should refer to ICAO Annex 14 for design details, which are graphically depicted within Figs 5.4 and 5.5 and Tables 5.6 to 5.8 inclusive. The reader should refer in particular to ICAO Annex 14 Chapter 3, Physical

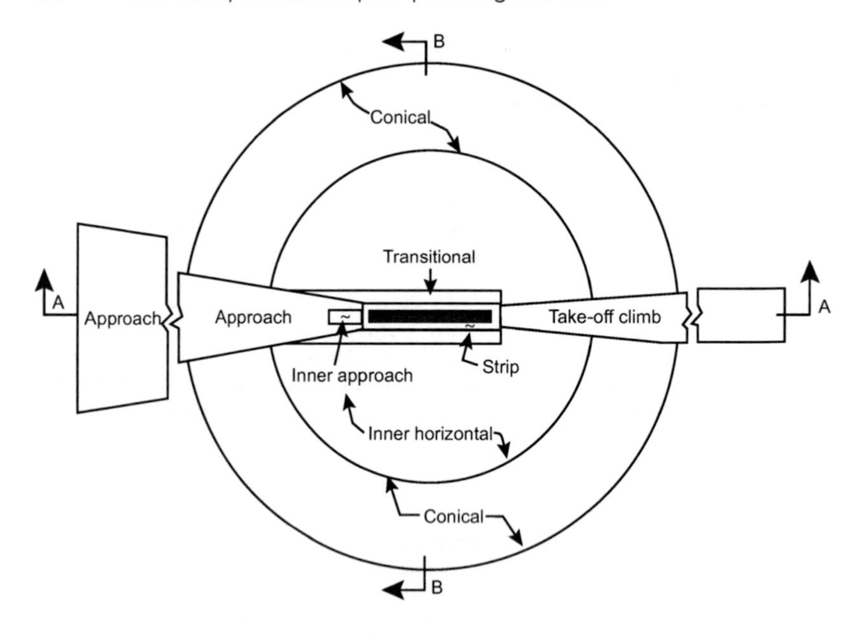

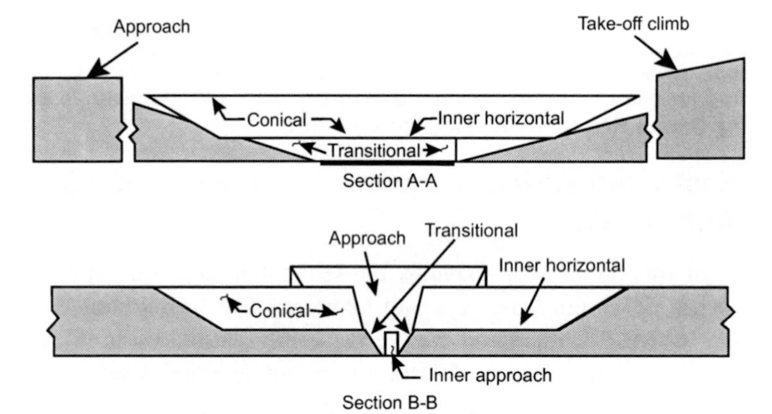

5.4 Obstacle limitation surfaces (ICAO figure 4-1; reproduced by kind permission of ICAO).

Characteristics, and Chapter 4, Obstacle Restriction and Removal. The obstacle limitation airspace surfaces surrounding an airport runway include:

- the horizontal section
- approach surface
- slope surfaces
- take-off climb surface
- transitional surface
- conical surface slope.

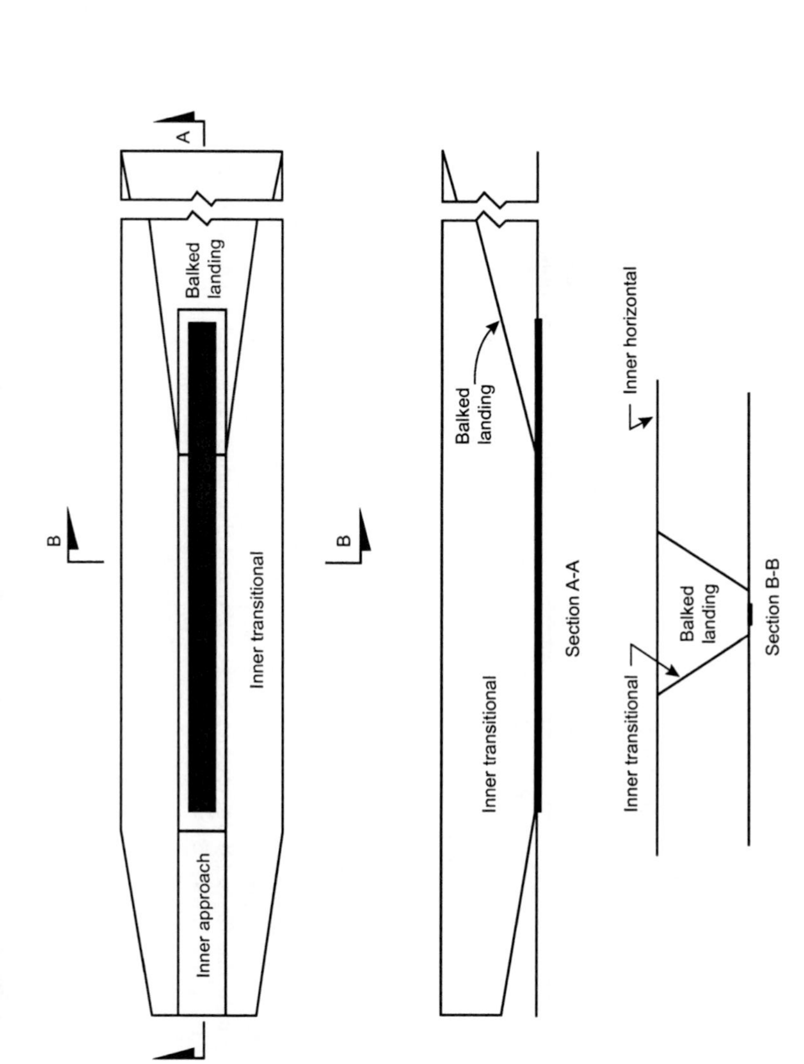

5.5 Inner approach, inner transitional and balked landing limitation surfaces (ICAO figure 4-2; reproduced by kind permission of ICAO).

Table 5.7 Taxiway minimum separation distances (ICAO table 3-1; reproduced with kind permission by ICAO)

Code letter	Distance between taxiway centreline and runway centreline (meters)								Taxiway centreline to taxiway centreline (metres)	Taxiway, other than aircraft stand taxilane, centreline to object (metres)	Aircraft stand taxilane,centreline to object (metres)
	Instrument runways code number				Non-instrument runways code number						
	1	2	3	4	1	2	3	4			
(1)	(2)	(3)	(4)	(5)	(6)	(7)	(8)	(9)	(10)	(11)	(12)
A	82.5	82.5	—	—	37.5	47.5	—	—	23.75	16.25	12
B	87	87	—	—	42	52	—	—	33.5	21.5	16.5
C	—	—	168	—	—	—	93	—	44	26	24.5
D	—	—	176	176	—	—	101	101	66.5	40.5	36
E	—	—	—	182.5	—	—	—	107.5	80	47.5	42.5
F	—	—	—	190	—	—	—	115	97.5	57.5	50.5

Note 1. The separation distances shown in columns (2) to (9) represent ordinary combination of runways and taxiways. The basis for development of these distances is given in the *Aerodrome Design Reference Manual*, Part 2.
Note 2. The distances in columns (2) to (9) do not guarantee sufficient clearance behind a holding aeroplane to permit the passing of another aeroplane on a parallel taxiway. See the *Aerodrome Design Reference Manual*, Part 2.

Table 5.8 Minimum distance from runway centreline to holding positions (ICAO table 3-2; reproduced with kind permission by ICAO)

Type of runway	Code number			
	1	2	3	4
Non-instrument	30 m	40 m	75 m	75 m
Non-precision approach	40 m	40 m	75 m	75 m
Precision approach category I	60 m†	60 m†	90 m*†	90 m*†‡
Precision approach categories II and III	—	—	90 m*†	90 m*†‡
Take-off runway	30 m	40 m	75 m	75 m

* If a holding bay, the runway-holding position or road-holding position is at a lower elevation compared to the threshold and the distance may be decreased 5 m for every metre the bay or holding position is lower than the threshold, contingent upon not infringing the inner transitional surface.
†This distance may need to be increased to avoid interference with radio navigation aids, particularly the glide path and localiser facilities. Information on critical and sensitive areas of instrument landing system (ILS) and microwave landing system (MLS) is contained in Annex 10, Volume I, Attachments C and G respectively (see also 3.12.6).
Note 1. The distance of 90 m for code number 3 or 4 is based on an aircraft with a tail height of 20 m, a distance from the nose to the highest part of the tail of 52.7 m and a nose height of 10 m holding at an angle of 45° or more with respect to the runway centreline, being clear of the obstacle-free zone and not accountable for the calculation of OCA/H.
Note 2. The distance of 60 m for code number 2 is based on an aircraft with a tail height of 8 m, a distance from the nose to the highest part of the tail of 24.6 m and a nose height of 5.2 m holding at an angle of 45° or more with respect to the runway centreline, being clear of the obstacle-free zone.
‡Where the code letter is F, this distance should be 107.5 m.
Note. The distance of 107.5 m for code number 4 where the code letter is F is based on an aircraft with a tail height of 24 m, a distance from the nose to the highest part of the tail of 62.2 m and a nose height of 10 m holding at an angle of 45° or more with respect to the runway centreline, being clear of the obstacle-free zone.

It is obviously possible to have various combinations of runway lengths, orientations and also technical landing instrumentation status. All of these parameters have the ability to change the obstacle limitation surfaces considerably. It is essential that the airport planner understands the operational restrictions that can exist often as a result of tower blocks and residential dwellings. By reverse engineering it is possible to determine the maximum capability for a new runway, taking into account the restrictive surfaces that currently exist. Commercial decisions can then be made by the designer/developer to decide if it is more appropriate either to accept the restrictions or actively look to redevelop/redesign the obstacle surfaces. The latter can often involve the costly purchase of new land and redevelopment of it to ensure that the obstacles or restrictions no longer exist. It is these types of very difficult and expensive decisions that will make it very important to have high-quality believable forecasts, which will enable the business case decisions to be made appropriately. Reference should be made, though not limited to, Figs 5.4 and 5.5 and Tables 5.6 to 5.8 inclusive.

Figures 5.4 and 5.5 showthe cross-sectional and plan position respectively of the transitional surface relative to the runway strip. It is important to understand that there are likely to be situations where the terminal and pier infrastructure, navigational aids and control tower(s), etc., could need to penetrate the transitional surface. This will require an appropriate operational justification and safety case for discussion and agreement from the national body authorising aircraft movements. In the UK this would be the Civil Aviation Authority (CAA) – National Air Transport Services (NATS) division and CAA Safety Regulation Group (SRG).

ICAO Annex 14 Chapter 3, entitled Physical Characteristics, specifies the runway length, runway width and taxiway design characteristics, which will enable the airport reference coding to be determined. Chapter 5 of this Annex specifies the visual aids and runway and taxiway marking systems, which in turn would be determined during the later stages of the scheme design.

Once it has been decided, with reference to Chapter 2 of this manual (see step 3, Understand the runway configurations options), which broad runway configuration to adopt, it will then be necessary to understand what runway length, width and precise position will be needed. The airport planner should refer to the forecast booklet to determine the forecasted aircraft to be accommodated along with aircraft movement frequency. The runway length and width will be determined by the aircraft code letters to be accommodated. The aeroplane reference field length for the runway shall be determined by the wing span (engine position essentially) and main outer landing gear wheel span (see Table 5.6).

The separation distances between parallel runways can be determined by referring to ICAO Annex 14 Chapter 3 Clause 3.1.10. Taxiway separation distances can be determined by referring to Table 5.7 and by referring to ICAO Annex 14 Chapter 3 Clause 3.9. Taxiway separation distances will be essentially determined by the aircraft code letters to be accommodated. The width of the runways/runway strips can be determined by referring to ICAO Annex 14 Chapter 3 Clause 3.1.9 and Clause 3.4.

The position of RETs and RATs should be considered at the masterplanning stage. The precise location will need to be defined during the scheme design stage, though with due consideration to runway slope, aircraft code sizes, aircraft deceleration performances and runway movement frequency. The RET and RAT infrastructures can be extremely expensive to provide, so it will be necessary first to model the operational and corresponding financial benefits to the airport by including them.

6 Design for airport security

Abstract: This chapter discusses design for airport security. It reviews threats such as hijacking and sabotage of aircraft as well as sabotage of airport facilities. After reviewing legislative requirements, it discusses design options for terminals, piers and perimeters.

Key words: airport security, hijacking, terminals, piers, airport perimeter.

6.1 Threats to aviation

Threats to aviation have been around for decades. Instances of terrorism date back to the 1930s. The common perception is that terrorism is the main threat, although most airports will never experience an act of terrorism. The main threats come from criminal activity, which often can be the result of dishonest passengers, criminal gangs, disgruntled staff, etc. The frequency of these criminal acts can be daily at large airports. Acts of terrorism are designed by the perpetrator to be high profile, intended to cause maximum disruption, fear and financial impact. Airports, no matter what size, should understand what the threats are. Those responsible should design airports and operational processes to counter these threats. This is an internationally mandated requirement stipulated in the ICAO Annex 17 security manual.

Smaller airports have in the past been used by terrorist groups as the weaker link in the transfer passenger journey to enable them to gain access to larger higher-profile transferring locations. Enforcement of enhanced Annex 17 requirements and the introduction and enforcement of post-9/11 FAA TSA legislation and IATA recommended practice has dramatically improved these historically weak links. Even the smallest of airports in the most remote location which has an in-line transferring function to a major international airport will need to have sophisticated hold baggage screening and passenger and staff screening processing in place. The capability of this equipment, which is continually improving, is extremely impressive and has undoubtedly acted as a deterrent to the would-be 9/11 or Pan Am 103 type of terrorist.

It is important that the potential threats are constantly (daily) reviewed and the teams and processes that are in place are adapted to these changing needs. These processes can be expensive to administer, but with effective design these types of threats can be reduced in impact and operational cost. This chapter aims to highlight what types of threats exist and how good design can minimise the impacts if these acts are carried out. The threats to aviation come from the following main areas:

- hijacking of aircraft
- sabotage of aircraft
- sabotage of airports.

These are discussed in the following sections.

6.2 Hijacking of aircraft

The first recorded incident of hijacking of an aircraft occurred in Peru in 1931. A second recorded incident occurred in 1947. A series of minor incidents occurred until 1966 when the Tri-continental Communist Congress was staged in Cuba, which resulted in a major hijacking incident, setting the standard by which future hijacking events were measured, leading up to the well-known hijacking events of 9/11. After 1969 there was a surge in incidents until 1973, when international terrorist groups regularly used the technique. The lowest global annual rate was 17 incidents per annum recorded in 1986 until around the mid 1990s. Approximately 75% of incidents involve scheduled commercial airlines. The 9/11 New York/Washington attacks were different in that the hijackers were arguably trained pilots who used the aircraft as weapons. There was no intention to use the passengers to barter for a set of political demands. The objective of the hijacking in this case was to attack the heartland of America, cause maximum death and destruction and dramatically raise the profile of the terrorist group.

Hijacking has been with us for decades and is not likely to go away unless measures are put in place to combat this risk. Facilities and processes can be put in place to help mitigate this threat, some of which the airport planner can influence when designing the new airport. These include:

- introduction of passenger and staff screening
- hand baggage screening – conventional – EDS – CT X-ray screening
- passenger screening archway metal detector (AMD) – particle analysis – full-body X-ray

- introduction of enforced restricted zones (airside) and access controlled areas
- introductionofhigh-resolution24/7CCTVandthermalimagingcameras
- below-ground pressure sensors for intruder detection.

These can complement operational improvements such as:

- comprehensive staff security character profiling
- aircraft cockpit locked doors
- air marshals.

6.3 Sabotage of aircraft

Since 1945 incidents of sabotage on aircraft have occurred frequently. Sabotage of aircraft occurs both on the ground and in the air. There have been in excess of 2500 people killed due to aircraft sabotage. Examples include Air India B747 (1985) and Pan Am Lockerbie (1988), where 257 passengers were killed. Sabotage of aircraft on the ground has also occurred. Three aircraft were destroyed at Colombo International Airport Sri Lanka (1986). In excess of 7 kg of high explosives were estimated to have been used, resulting in 16 passengers being killed.

The airport planner can reduce the threat of this scenario by ensuring that all staff are screened through staff screening points, which are strategically located across the airport. Aprons should be provided with good lighting and should be adequately provided with clear sight lines from airfield operational positions and control towers. Restricted zone areas should be clearly demarked and access systems linked to systems that allow these zones to be controlled effectively. The design of aprons should be covered by high-resolution 24/7 CCTV and thermal imaging cameras as necessary. From an operational perspective, comprehensive and effective staff security character profiling should be carried out. Security patrols should be employed regularly and patrol behaviour and frequency should not be predictable. Staff access to apron areas should be limited to the absolute minimum. Aircraft should be kept locked (tamper locks) when not in use.

6.4 Sabotage of airports

The objective of sabotage at airports is arguably to cause disruption, death and damage to property and to raise the profile of the perpetrator. Often politically and religiously motivated terrorists seek to maximise damage and injury. Environmental terrorist/pressure groups are also resorting to sabotage. In winter 2008 Stansted Airport in the UK witnessed a breach of the perimeter fence and restricted zone. As airports continue to expand, this type of activity could increase. Most attacks occur in public areas,

6.1 Failed attempt to blow up the Glasgow Airport check-in building using a car bomb.

although airport installations are being attacked more frequently. The airport designer should look to make it as difficult as possible to gain unauthorised access to runways, the airfield lighting and approach guidance systems, sensitive landside terminal infrastructure and railhead infrastructure. Examples of violent sabotage and attacks of airports include:

- the Rome and Vienna simultaneous attacks of 1985;
- Israeli passengers attacked with machine guns while in a check-in queue (Germany);
- the attempt to ram a car loaded with explosives into the check-in building at Glasgow Airport (UK; see Fig. 6.1).

6.5 Legislation: international, European and domestic obligations

The legislation pertaining to airport design function is formed of three sections:

- international legislation
- Continental legislation
- domestic legislation.

With respect to international legislation, the highest global standard is set by the legally mandated requirements of the International Civil Aviation Organisation (ICAO-OACI). The documents of main interest to the airport planner and security systems designer for the airport are:

- Annex 14, Volume 1, Aerodromes and Volume 2, Aerodrome Design and Operations
- Annex 17, Security.

ICAO-OACI sets *International Standards and Recommended Practices* that participating states must follow. ICAO-OACI is a subset of the United Nations, with the prime objective to ensure that international civil aviation is developed safely and in an orderly manner.

Annex 14 contents include:

- aerodrome data category definition
- runway and taxiway design characteristics
- visual aids, markings, lighting, signs – obstacle markings
- electrical systems
- aerodrome operational services
- aerodrome maintenance.

Annex 17 contents include:

- general principles of aviation security
- organisation of security procedures
- preventive security measures
- management of response to acts of unlawful interference.

With regards to Continental legislation, these include the European arm of ICAO-OACI known as the European Civil Aviation Conference (ECAC). The ECAC Document 30, Restricted Access, is a document of prime interest to airport designers and security system designers. European countries and designers and airport operators must abide by standards and recommendations set in ECAC Document 30. Airport planners and security system designers working on airport developments within the European Union should also review the following legislation: Regulation (EC) No. 300/2008 of the European Parliament and of the Council of 11 March 2008 on common rules in the field of civil aviation security and repealing Regulation (EC) No. 2320/2002.

Domestic legislation varies enormously. Broadly, most countries with major airports follow the recommended practices defined by ICAO-OACI and expand upon the generic ICAO-OACI documentation to provide local interpretation and definition suitable to provide safe and efficient operations in the country in question. The ICAO-OACI and ECAC recommended practices are international best practice aspirations and include statements that do not explicitly define the technology and procedures necessary to enable their intent to be adhered to. It is usually the local domestic legislation that does this. Good examples of this are:

- Canadian Air Transport Security Authority (CATSA) (Canada)
- FAA/TSA (United States)
- DfT (United Kingdom)
- CAA CAP 168.

6.6 ICAO Annex 17

The following clauses have been replicated from ICAO Annex 17 with kind permission of the ICAO. Following each ICAO recommendation there is a statement from the author designed to aid the airport planner or security specialist in understanding the intent of the ICAO recommendation.

Chapter 3. Physical characteristics
3.13 Isolated aircraft parking position
3.13.1 An isolated aircraft parking position shall be designated or the aerodrome control tower shall be advised of an area or areas suitable for the parking of an aircraft which is known or believed to be the subject of unlawful interference, or which for other reasons needs isolation from normal aerodrome activities.

Author's design statement. Isolated parking areas can be surrounded by an intruder detection system, e.g. below-ground sensors, also visual surveillance systems such as CCTV and thermal imaging equipment. It is also possible to locate concealed emergency squad team holding and control post areas and also strategically positioned and concealed sniper posts. Often this is best done using earth banks.

3.13.2 Recommendation — The isolated aircraft parking position should be located at the maximum distance practicable and in any case never less than 100 m from other parking positions, buildings or public areas, etc. Care should be taken to ensure that the position is not located over underground utilities such as gas and aviation fuel and, to the extent feasible, electrical or communication cables

Author's design statement. These isolated aircraft parking stands should be positioned to ensure that the airport can remain operational while the unlawful activity is resolved. This can be achieved by providing physical barriers between the isolated parking stand and those of the fully operational taxiway and runway.

Chapter 2. Entry and departure of aircraft
A. General
2.2 Contracting States shall make provision whereby procedures for the

clearance of aircraft, including those normally applied for aviation security purposes, as well as those appropriate for narcotics control, will be applied and carried out in such a manner as to retain the advantage of speed inherent in air transport.
Note.— With respect to application of aviation security measures, attention is drawn to Annex 17 and to the ICAO Security Manual.

Author's design statement. The airport planner needs to ensure that the operational processes are slick and effective and that all reasonable attempts are made to ensure that risk scenarios for the particular airport are understood and mitigated accordingly using appropriate technology.

Chapter 3. Entry and departure of persons and their baggage
A. General
3.2 Contracting States shall make provision whereby the procedures for clearance of persons travelling by air, including those normally applied for aviation security purposes, as well as those appropriate for narcotics control, will be applied and carried out in such a manner as to retain the advantage of speed inherent in air transport.
Note.— With respect to application of aviation security measures, attention is drawn to Annex 17 and to the ICAO Security Manual.

Author's design statement. In practical terms this means that facilities and processes should be put in place which should allow staff and their belongings that proceed through the restricted zone boundary to be 100% security searched. Goods could be screened using conventional or smart EDS type screening machines with particle analysis machines used for trace detection subject to local law mandates. Staff could be screened using automatic metal detection walk-through machines or full-body X-rays supplemented again with particle analysis machines, subject to local law mandates.

C. Departure requirements and procedures
3.33 Contracting States shall, in conformity with their respective regulations, endeavour to reduce the documentation required to be produced by passengers departing from their territories to a valid passport or other acceptable form of identity document.
Note.— It is understood that such documentation should include a valid visa if required.

Author's design statement. The objective of this statement is to make airports and airlines employ the passport document as the major entry and exit document used where possible. It should be noted that many passports still exist that are valid but do not contain the necessary biometric data to provide sufficient confidence that the passport and the person presenting the passport are the same person. As machine readable biometric data become securely embedded into passports the need for countries to impose further document controls or protocols will start to reduce.

3.34 Contracting States shall not require the presentation or inspection of baggage of passengers departing from their territory, except for aviation security measures or in special circumstances.
Note.— This provision is not intended to prevent the application of appropriate narcotics control measures and specific customs control where required.

Author's design statement. Basically the airport is required to screen all baggage for safety and security reasons. The recommendation states that separate narcotic controls can be put in place to meet Customs control requirements. The Customs control authorities may use special narcotics screening machines or procedures that involve K9 operations.

Chapter 8. Equipment and installations
8.1 Secondary power supply
General Application
8.1.1 Recommendation.— A secondary power supply should be provided, capable of supplying the power requirements of at least the aerodrome facilities listed below:
e) Essential security lighting, if provided in accordance with 8.5;

Author's design statement. This statement is a key security requirement for an aerodrome. A separate arguably hot standby back-up power supply is an essential requirement to be fitted. The secondary power supply could be positioned in a separate location and both primary and secondary power incomer and airport distribution lines located inside the restricted zone.

8.4 Fencing Application
8.4.2 Recommendation.— A fence or other suitable barrier should be provided on an aerodrome to deter the inadvertent or premeditated access of an unauthorized person onto a non-public area of the aerodrome.
Note 1.— This is intended to include the barring of sewers, ducts, tunnels, etc., where necessary to prevent access.
Note 2.— Special measures may be required to prevent the access of an unauthorized person to runways or taxiways which overpass public roads.
8.4.3 Recommendation.— Suitable means of protection should be provided to deter the inadvertent or premeditated access of unauthorized persons into ground installations and facilities essential for the safety of civil aviation located off the aerodrome.

Author's design statement. The complete perimeter of the airport should be protected by a security graded fence plus in certain locations additional intruder detection systems. This sounds straightforward, but the perimeter of most Code C plus serving airports can be in the region of 10 km. The perimeter will interface with highly populated areas, remote areas, rivers, passenger and staff access zones and cargo and airfield servicing vehicles – to name but a few of the many interfaces. The airport planner should work with the local security teams to understand the terrain of the perimeter, its vulnerabilities and local and national threat intelligence. Armed with this knowledge the airport planner should look to plan in a suitable fencing system, camera systems and below-surface intruder detection systems.

> *8.4.4 Recommendation.— The fence or barrier should be located so as to separate the movement area and other facilities or zones on the aerodrome vital to the safe operation of aircraft from areas open to public access.*

Author's design statement. This clause requires the airport planner to break up the zones of the airport into logical areas, e.g. apron, control tower, baggage hall, fire ground, terminal and pier areas. In some countries this is done by using colour coded restricted access control passes or smart radio frequency locator passes. The smart passes actively monitor staff locations on the airport. If there is a person on, say, the airfield with no pass or more commonly with the wrong access permission, the central control room is notified immediately.

> *8.4.5 Recommendation.— When greater security is thought necessary, a cleared area should be provided on both sides of the fence or barrier to facilitate the work of patrols and to make trespassing more difficult. Consideration should be given to the provision of a perimeter road inside the aerodrome fencing for the use of both maintenance personnel and security patrols.*
>
> *8.5 Security lighting*
>
> *Recommendation.— At an aerodrome where it is deemed desirable for security reasons, a fence or other barrier provided for the protection of international civil aviation and its facilities should be illuminated at a minimum essential level. Consideration should be given to locating lights so that the ground area on both sides of the fence or barrier, particularly at access points, is illuminated.*

Author's design statement. The perimeter fence security characteristics can be significantly enhanced if the fence is located with a clearway on both sides. This enables patrols to monitor the total perimeter and not be stopped by foliage, which often can obstruct clear sight lines. It is important that the location of the perimeter fence is not compromised by structures or terrain that could render it useless. This basic error has been evident at some major airports. Do not allow fences to be located where they are buried in local ditches, permitting intruders to simply jump over the top of the fence! Often these mistakes are the result of shoddy workmanship from contractors trying to cut costs. Perimeters should also be suitably illuminated with lux levels similar to that used in low-level lighting of car parks. Staff and vehicle access points should be very clearly illuminated, as should areas of sensitive security characteristics. It is the recommendation of the author that, on the landside zone of the fence, particularly where there is vehicular access, vehicle ditches are dug in. The ditch needs to be at least 1.5 m deep (B in Fig. 6.2) and up to 4 m wide (A). The slope of the ditch should be near vertical on the side closest to the fence and gradual on the opposing side. This induces the vehicle to drop into the ditch, even a 4 × 4 driven at speed.

6.7 Masterplan airport development considerations

Airport planners, architects and airport designers should regularly work with the local and national security intelligence support network to determine the ever-changing perceived and real threats that may exist at an airport. A structure should be put in place to enable regular dialogue between these parties so that designs are well informed. Once designers understand the risks and threats that exist they should then consider a balanced approach to mitigating these risks. It is impractical to mitigate all risks but it is possible to manage all risks to a reasonable level using a combination of smart design and appropriate operational processes.

The key to successful secure airport design is the implementation of designs that aid operational security processes. As an example, do not include balconies that are open to passengers in landside areas. This will reduce the amount of security patrols necessary to make this highly vulnerable area safer. Ensure that cargo areas are correctly located with clear lines of sight so that security patrols can view areas easily and quickly. These basic principles can be obtained by designers spending a day or so with the security patrols to understand their current frustrations and operational inefficiencies.

In principle the design team should look to:

- Gather intelligence – understand the threats.
- Assess the risk and threats – produce a design that looks to mitigate risks.
- Apply technology efficiently – use CFD/walk-through simulations to predict design performance.
- Operation – implement complimentary operational processes.
- Constantly review measures – stay one step ahead.
- Design to protect key airport points:
  - passenger-intensive areas

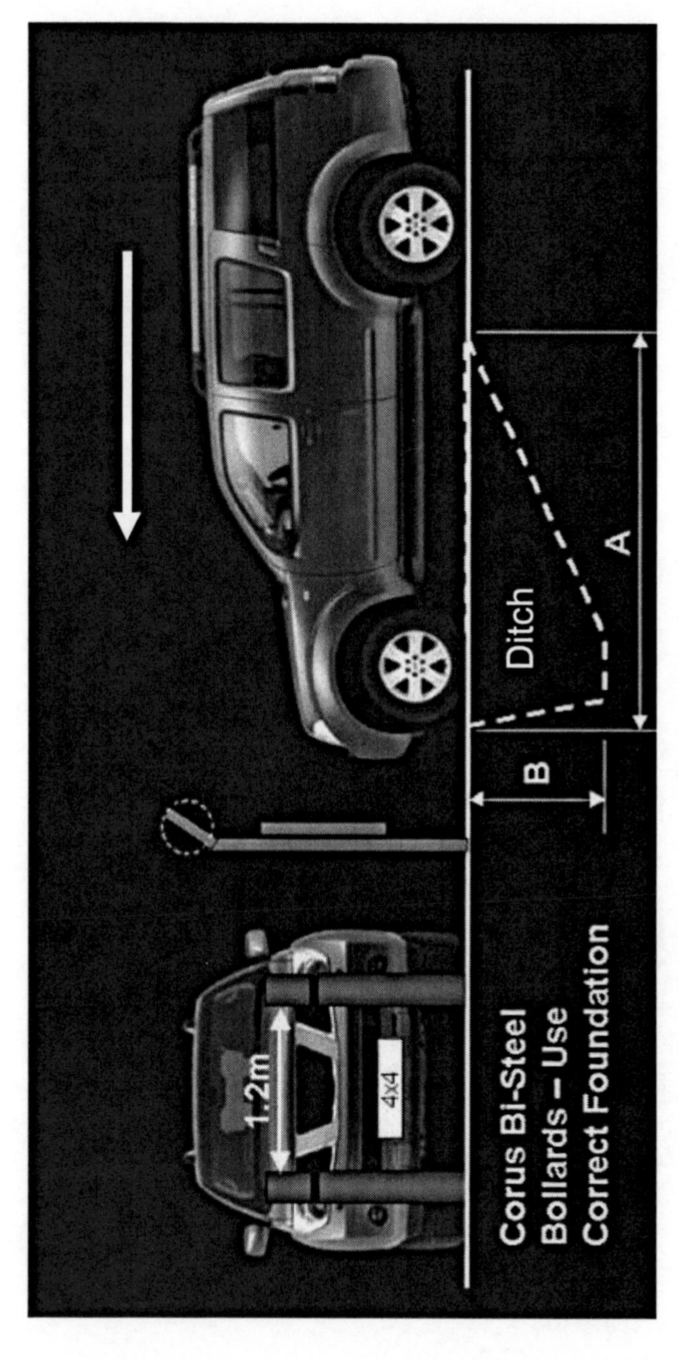

6.2 The use of ditches, fences and bollards for airport perimeter security.

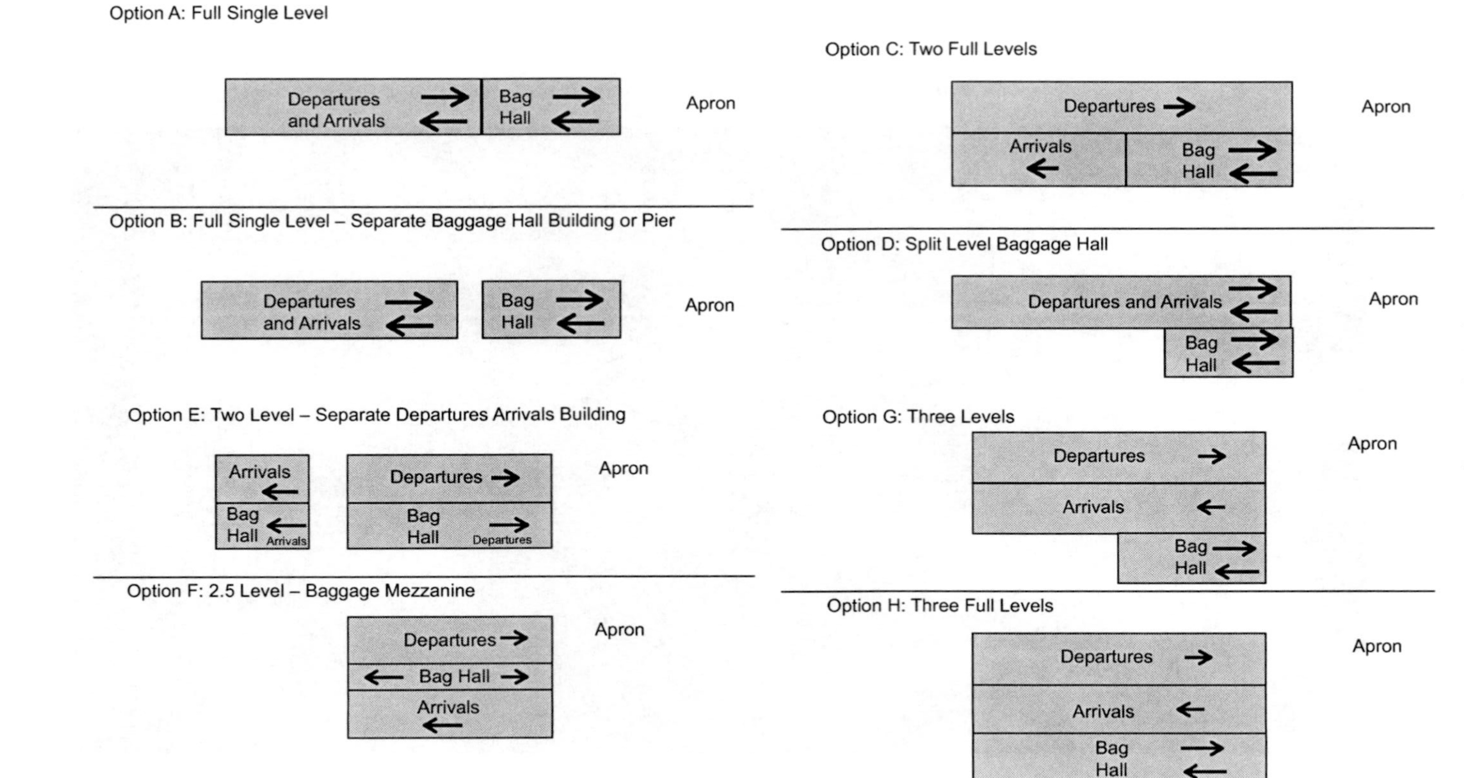

6.3 (a) Design options for terminals and piers/satellites: typical terminal cross-sections (continued on next page).

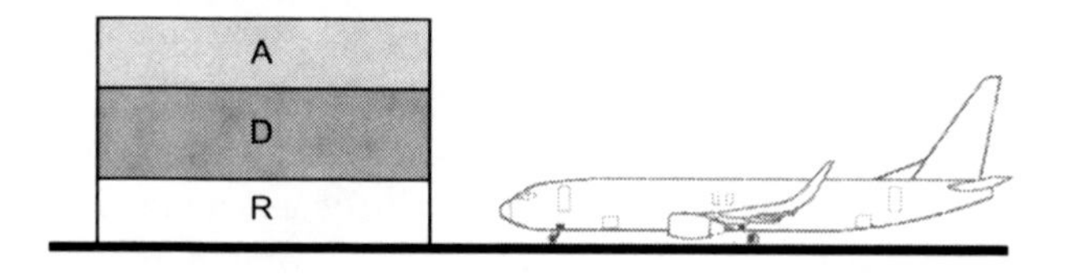

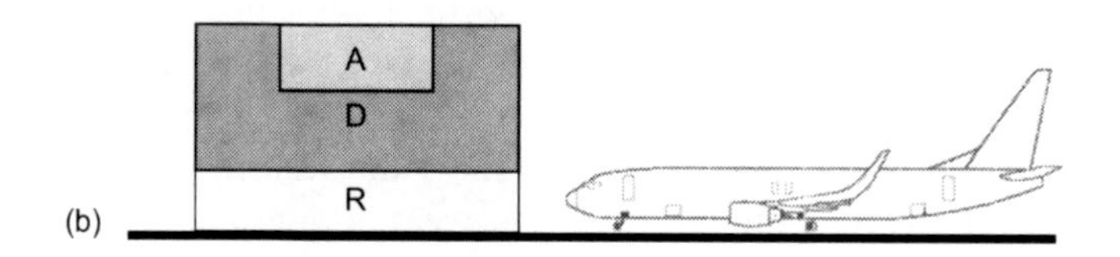

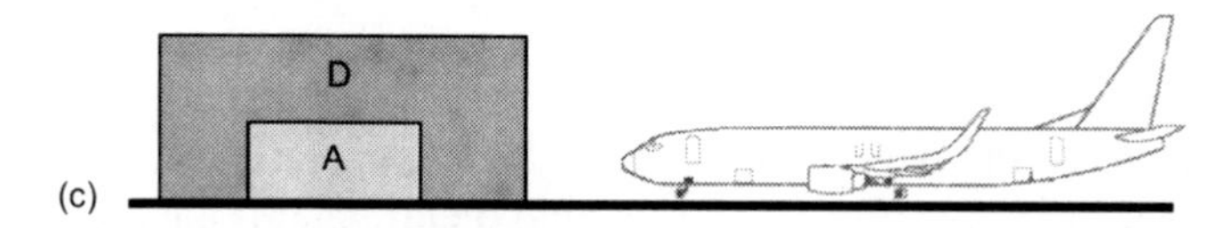

6.3 Design options for terminals and piers/satellites: (b) two and one-and-a-half level cross-section options for piers/satellites; (c) single-level option for piers/satellite (A = arrivals, D = departures, R = restricted zone; not to scale; other options exist).

- fire services
- power supplies
- fuel supplies
- sewage services
- communications processes.

6.8 Design options for terminals and piers

Figure 6.3(a) details a number of cross-section and baggage hall positioning options, A–H inclusive, that can be employed. Each of these options has different pros and cons, which are discussed below. Figures 6.3(b) and (c) show vertical design options for separating arrivals and departures.

Option A: full single level. This option includes a terminal building with a departures and arriving passenger function on a single building level physically connected to the baggage hall. While there are financial and flexibility advantages with the configuration there are equally major operational security disadvantages that can exist. For example, if there is a security alert inside any point of the building it is likely that the whole operation is disrupted. This could be either a suspect bag that cannot be reunited with its passenger or a passenger's bags within the baggage hall that has perhaps gone to level 4/5 of security screening. These scenarios to some extent can be designed out with clever design, but in the main do inherently cause more disruption due to security alerts.

Option B: full single level – separate baggage hall. The security performance is virtually the same as Option A described previously but Option B benefits from the separation of the baggage hall, which means passenger and baggage alerts do not disrupt each other.

Option C: two full levels. This option has the same security performance as Option A but the operational risk in the event of a serious security or fire risk on the lower levels will mean that the upper level is more likely to be more seriously affected when compared to Option A or Option B.

Option D: split-level baggage hall. This option has improved security performance when compared to base Option A. The extent of the baggage hall is such that in the event of a security alert in the baggage it is quite reasonable to conclude that the vast passenger space, which is not directly above the baggage hall, could be left operational. It should be noted that operational restrictions would dictate that if a baggage hall is under an alert, passengers on the upper level are not permitted to be directed to a space that could put them into danger.

Option E: two level separate departures and arrivals buildings. This option is quite common and an example includes the current (2010) format Terminal 3 operation at Heathrow. The main benefit is that the departures functions for passengers and their baggage are closely located together, allowing level 4 bags to be reunited more easily. A security alert in one building does not disrupt the operation of the adjacent but separate building. Obviously a baggage alert in one building affects the passenger option in the same building.

Option F: 2.5 level building. This option has the same security performance as Options A and C. It is quite a common format and actually is not that flexible to forecast change. This configuration is used when the land availability is small and is deemed more appropriate to contain the development area. A recent example of this is Terminal 5 at Heathrow, which is a multiple level building situated between two parallel runways.

Option G: three levels. This option has the same security characteristics as Option D.

Option H: three full levels. This option has similar security characteristics as Options C and F.

Table 6.1 Effects of explosions on building structural types

Distance of 5 kg high explosive device from building	Effect on steel-framed construction	Effect on load-bearing concrete structure
<5 m	Severe damage; likely local structural collapse	Complete structural collapse
>5 m < 10 m	Significant damage to façade	Major collapse
> 10 m < 15 m	Moderate damage to building façade	Structure not capable of being repaired
> 15 m < 20 m	Relatively minor damage to façade	Significant damage to façade that could be repaired
<20 m < 30 m	Superficial damage to façade; cladding and glazing system damaged by shock waves	

6.9 Impact of explosives on terminal infrastructure

Table 6.1 outlines the two types of basic building structure that commonly exist and then attempts to explain the effects that those structures will have when an explosive device is detonated in varying proximity to the building types. The typical explosive device results in primary and secondary fragments or debris being witnessed. The fragments in combination with the shock waves are what cause the damage and injury. When an explosive device is detonated the shock waves are unidirectional, i.e. shock waves expand in all directions at the same rate until they hit a surface. If the surface is soft, as in the case of general landscaping soil, the crater is larger than if the surface is hard and made of, say, concrete. The lateral shock waves from the explosion travel until they hit surfaces. As soon as a static surface is encountered by a shock wave, the immense energy is transferred into the static surface in fractions of a second, causing serious damage. If the structure is designed so as to deflect the shock wave, this can significantly reduce the damage.

Primary fragments are created by the device and its container. Secondary fragments are created by the blast wave, initially destroying static surfaces such as (but not limited to) glazing or masonry. The distance over which primary fragments can result in casualties is approximately twice that of secondary fragments.

All vehicles should be kept at least 50 m away from the façade of the terminal. Ideally, the forecourt roads should be at a lower level, creating a sloping ramp, which would act as a blast deflector from the primary shock wave. Modes of transport should be separated as denoted in the following list:

- Priority 1: rail
- Priority 2: bus and coach, including mid- and long-stay car park coaches
- Priority 3: taxi
- Priority 4: passenger cars (drop off, no dwell), arguably remove owing to security issues.

Suggested designs keeping these priorities in mind are shown in Fig 6.4.

6.10 Perimeter security

The perimeter of an airport is arguably the most difficult asset to defend. It is essential to understand the most vulnerable points of the perimeter, which will naturally vary from airport to airport and from month to month of operation. It is for this reason that regular local and national intelligence is gathered and assessed. Once the threats are understood appropriate infrastructure, systems and processes can then be put into place to mitigate these risks (see examples shown in Fig. 6.5). Where the airport planner is confronted with the task of defining a new airport perimeter, the use of this knowledge will help to set a perimeter which is easier and less complex and less expensive to defend its integrity. Areas that are hard to defend include: rivers, highly populated or frequented public areas, areas of forests, lake boundaries and areas that have high volumes of permitted traffic. Each presents unique challenges.

It is important that the fencing systems deployed are compatible with the terrain. The ICAO Document 9184 AN/902, *Airport Planning Manual*, recommendation is a useful reference for the use of such designed terrain. The European ISO specification is the basic minimum chain link recommendation, although this is really the entry level standard. Superior fencing systems exist such as pressed steel (vertical bars with various grades of steel quality and anti-corrosion) and/or welded mesh, of which each has its pros and cons. All fence lines will need to penetrate the soil level significantly. It is the author's recommendation that the fence media (chain link/pressed steel/welded mesh) line should extend no less than 1 m below the surface.

To aid the security integrity of the airport perimeter the following checklist items should be *observed*/considered by the airport planner:

- Gather intelligence from patrols and from national security threat agencies.
- Review, analyse gathered intelligence and put into place a rolling weekly/monthly reviewed threat mitigation plan (TMP).
- Subject to findings of the TMP, consider the use of thermal imaging, infrared and general high-resolution CCTV cameras and other intruder detection systems.

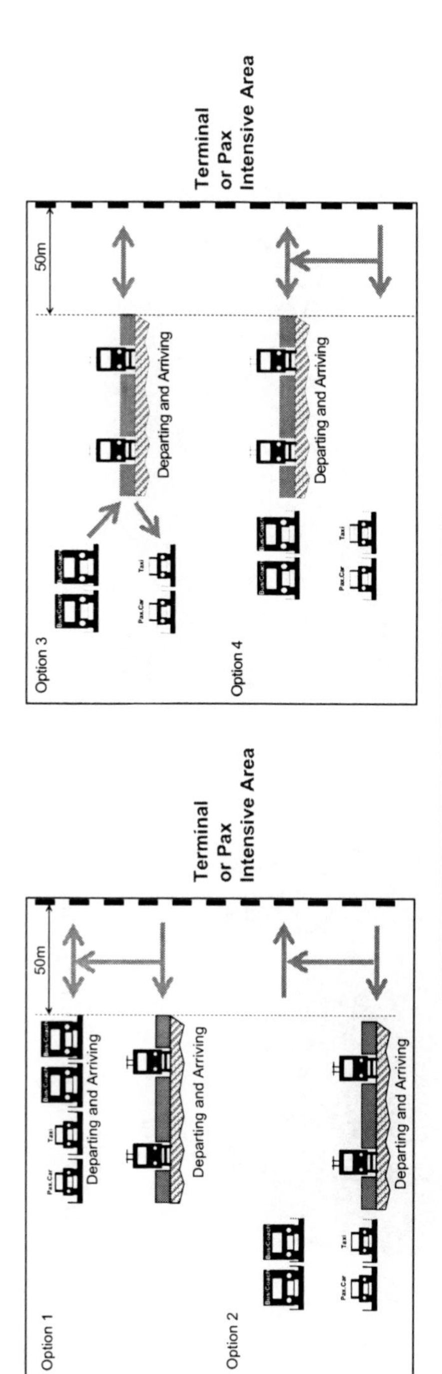

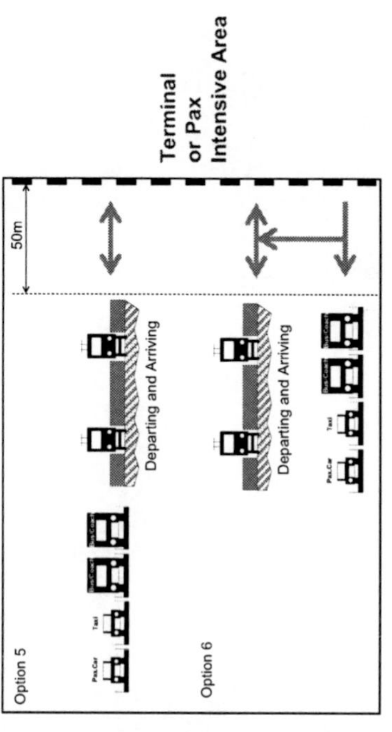

6.4 Option configurations for modes of transport close to a terminal façade.

6.5 Design options for perimeter security.

- Employ software and systems that detect new and common landside vehicle movement trends. Some of the most suspect vehicles have been cloned or commonly seen landside vehicles. Abnormal routes can be detected using vehicle transponders or number plate recognition. A number plate recognition system alone is not sufficient to rely completely upon in isolation.
- Exclude all vehicles from areas of high passenger density where possible.
- Have a process in place so that only permitted taxis are allowed on to the airport taxi ranks. Ensure that taxis are called on demand from more remote bulk taxi holding ranks, which are located on the airport perimeter.
- Ensure that terminal façades and forecourts are of a blast-proof design. The forecourt should be of an elevation which ensures that primary shock waves from, say, a vehicle bomb are not directed towards the terminal façade or towards areas of high passenger occupancy. Consider the use of blast deflectors.
- Make sure that the airport terrain aids security; e.g. do not install fence lines adjacent to the natural local mounds or hills. Ensure sensitive remote areas have landside ditches in front of the fence line of a suitable depth and profile, as described earlier.
- Ensure that perimeter security patrols are operationally deployed. The frequency of the patrols should not be predictable.
- Ensure that the perimeter is maintained adequately. This includes maintenance to the infrastructure and systems used and foliage maintenance. This will also include removal of debris from rivers and lakes that intersect with the airport perimeter. Ensure that a clear 3 m separation from the fence line each side of the perimeter is maintained at all times.

7
Case studies in airport planning

Abstract: This chapter includes a number of sample research questionnaires completed for five airports. The questionnaire format is designed to highlight likely planning and development issues affecting airports. The questionnaires illustrate what to look for and allow a comparison with real data from the airports concerned.

Key words: airport research report, Chisinau E. E. International Airport, Cologne Bonn Airport, Adelaide Airport, Luton Airport, Frankfurt Hahn Airport.

7.1 Introduction

Chapter 7 provides completed research questionnaires for five case study airports. There are two formats of questionnaire used. The larger more complex questionnaire has been provided for Chisinau S.E. International Airport in Moldova, for Cologne Bonn Airport in Germany and for Adelaide Airport in Australia. The fourth and fifth less complex airport questionnaires detail Luton Airport and Frankfurt Hahn Airport. The author has historically found both these formats to be useful. The larger questionnaire covers more attributes and is very comprehensive and useful for detailed research purposes. Many useful facts and figures can be obtained for comparison purposes if the larger questionnaire is used. The smaller questionnaire is useful for capturing initial data, which when known will allow the designer to understand if an airport is appropriate to select for further research using the full questionnaire.

The author has provided the full original data of the selected airports to allow the reader to be able to compare any dataset between the airports shown. This technique is effectively a benchmarking exercise. In the past the author has used the larger more complex questionnaire to compare the following data ranges for multiple airports:

(i) Field 1.0 Airport Capacity versus Field 2.1: What is the technology split of check-in?
(ii) Field 1.0 Airport Capacity versus Field 2.5: Does the baggage handling system incorporate automatic or manual baggage sortation?
(iii) Field 1.0 Airport Capacity versus Field 5.1: What % proportion of the terminal building area is dedicated to airside and landside retail?
(iv) Field 1.0 Airport Capacity versus Field 13.4: What are the runway length(s)?
(v) Field 1.0 Airport Capacity versus Field 13.5: What is the runway operating mode (segregated/compass mixed)?
(vi) Field 13.9: What is cargo throughput (tonnes/annum) versus Field 13.10: What is cargo shed size?
(vii) Field 13.9: What is cargo throughput (tonnes/annum) versus Field 13.11: What is the cargo apron (stand) area?

There are many combinations of comparison that can be used to obtain the correct benchmarking to ensure that a new airport facility or infrastructure design is fit for purpose, appropriately sized and fitted out correctly and within the ball park of comparative airports.

7.2 Airport research report: Chisinau S. E. International Airport, Moldova

Airport name/country

Chisinau S.E. International Airport, Moldova

Report by

Alla Tubari, with supplementary data collected by the author

Date

18 March 2009

1 Airport capacity/demand/miscellaneous issues

1.1 What is the current MPPA (2009/10)?

930 000 pax/940 000 pax

1.2 What is the MPPA split between departing passengers and arriving passengers (%)?

Departures	49.6%
Arrivals	50.4%
Total	100%

1.3 What is the forecasted MPPA for the next 5 years?

2010; 930 000 pax
2011; 940 000 pax
2012; 960 000 pax
2013; 1 000 000 pax
2014; 1 100 000 pax

1.4 What level of future infrastructure is planned in order to accomplish this growth?

Requires modernisation

1.5 Do you have published service level space standards (space per passenger)?

N/A

2 Check-in

Passenger operations

2.1 What is the technology split of check-in?

	Now	5 years' time
Conventional	12 desks	N/A desks
CUSS	0 kiosks	N/A kiosks
Branded self-service	0 kiosks	N/A kiosks
Online	0 % pax	N/A % pax

2.2 What is the designed space per passenger in this zone (m^2/pax)?

1.6 m^2/pax

Baggage operations

2.3 How many dedicated 'Baggage Drop' positions are used with common user self-service (CUSS)/self-service check-in (SSCI)?

N/A

2.4 How many oversized baggage drops are provided?

1

2.5 Does the baggage handling system incorporate automatic or manual baggage sortation? Is there an airline baggage sortation technology preference?

Manual baggage sortation used

2.6 What is the floor area of the baggage handling system?

900 m^2

3 Security

3.1 Do you have:

- **Centralised security only?**

Yes only centralised

- **Gate security only?**

N/A

- **Both centralised and gate security?**

N/A

3.2 Are security costs subsidised by the Government?

Partially N/A

3.3 Do you process hand and hold baggage through central passenger security X-ray machines?

Yes

3.4 What is the processing rate of passengers through the security functions?

1 min

3.5 What is the designed space per passenger in this zone (m^2/pax)?

1.5 m^2/pax

4 Departures immigration

4.1 Is there a requirement for departures immigration?

Standard procedures

4.2 On average how long does it take to process EU/non-EU passengers through the departures immigration process?

American	pax/min	2 pax/min
Non-American	pax/min	2 pax/min
EU	pax/min	2 pax/min
Non-EU	pax/min	2 pax/min

4.3 What is the designed space per passenger in this zone (m^2/pax)?

1.5 m^2/pax

5 Departures retail lounges

5.1 What % proportion of the terminal building area is dedicated to:

- **Airside retail?**

  7.2 % N/A m^2

- **Landside retail?**

  5.1 % N/A m^2

5.2 Is there retail within the piers/satellites?

N/A

6 Passenger delivery

6.1 What is the maximum airside walking distance witnessed for passengers?

N/A

6.2 What proportion (%) of passengers use disabled facilities (any special facilities)?

N/A

6.3 What proportion (%) of stands/gates are pier/satellite served?

N/A

6.4 Is there a tracked transit system (mass people movement system)? If yes, what is its function? – landside terminal transfers or airside terminal to pier connections?

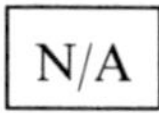

7 Arrivals immigration and transfers

7.1 How many arrivals immigration desks are used?

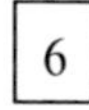

7.2 Typically how long does it take American/non-American/EU/non-EU passengers to progress through arrivals immigration?

American	2 min
Non-American	2 min
EU	2 min
Non-EU	2 min

7.3 Are biometric systems used/planned to be needed at immigration?

Yes

7.3.1 How long does it take to process passengers using this technology?

American	1 min
Non-American	1 min
EU	1 min
Non-EU	1 min

7.4 Is there an airside transfer function and what percentages of arriving traffic are associated with transfers?

- **Airside transfer**

  0 %

- **Landside transfer**

  100 %

7.5 Is the airside transfer function only for full-service carriers or do low-cost carriers use them?

Only for full-service carriers

7.6 How many security search archways and screening machines are used?

2 units

7.7 What is the designed space per passenger in this zone (m^2/pax)?

2 m^2/pax

8 Arrivals baggage reclaims

8.1 How many baggage reclaims are used?

1

8.2 What is the designed space per passenger in this area (m^2/pax)?

1.6 m^2/pax

8.3 Any special features?

N/A

9 Arrivals customs facilities

9.1 What is the area of the customs facilities?

50 m^2

9.2 What is the designed space per passenger in this area (m^2)?

1.5 m^2/pax

10 Energy management systems

10.1 Are there energy management systems in place?

- **Lighting systems?**

Yes

- **Baggage systems?**

Yes

- **Passenger travelators (moving walkways)?**

- **Other?**

10.2 What is the energy consumption/generation strategy for the airport?

6 925 600 kW/ hour for main system and alternative system

11 Waste management systems

11.1 What waste management strategies are in place? Is waste broken down into categories?

No

11.2 Are there waste compactors on site?

12 Fire strategy

12.1 How does the fire evacuation strategy for the main terminal building work?

N/A

12.2 Do the building level changes experienced by the passengers create any problems for building evacuation?

Yes

13 Apron questions

13.1 Please attach a layout of the airport/apron showing runways and taxiway positions.

Available ☑	Not available ☐

13.2 Explain the general functionality of the airfield.

Primary runway	(08/26) serves take-offs and landings
Secondary runway	(N/A) serves No
Other No..	

13.3 For what wing span/ICAO Code has the airfield been safeguarded?

For aircraft with a C code letter

13.4 What are the runway length(s)/orientations?

Runway (08/26) length =	3590	m
Runway ( ) length =		m
Runway ( ) length =		m
Runway ( ) length =		m

13.5 What is the runway operating mode (segregated/compass mixed mode, etc.)?

MM – single

13.6 What is the average and maximum taxi distance from the runway to stand?

From 600 to 1200 m

13.7 What is the average aircraft turn-around time?

Low-cost carrier	min (B737-800 say)	N/A
Full-service carrier	min (B737-800 say)	55 min

13.7.1 How is the passenger boarding managed for low-cost flights?

N/A – walk to gate

13.8 What is the passenger apron size?

40 000 m^2

13.9 What is the cargo throughput (tonnes/annum)?

3.000 tonnes/annum

13.10 What is the cargo shed size?

214 m^2

13.11 What is the cargo apron (stand) area?

10 000 m^2

13.12 Are the low-cost stands served with airbridges?

N/A – walk to gate

14 Passenger experience

14.1 Are there any customer satisfaction surveys carried out across the airport? If 'yes', could we have a copy of this material?

Yes – see Appendix 1

14.2 For each statement listed below please tick one box that best describes the effectiveness of the airport function when viewed from a passenger's perspective.

	Very good	Good	Average	Poor
Ease of entering the airport complex by train	☐	☐	☐	☐
There are no trains to the airport				
Ease of entering the airport complex by car	☑	☐	☐	☐
Ease of entering the airport complex by public bus	☐	☑	☐	☐
Effectiveness of check-in processes	☐	☑	☐	☐
Suitability and size of *landside* retail offer	☐	☐	☐	☑
Speed and effectiveness of security processes	☐	☐	☑	☐
Suitability and size of *airside* primary retail offer	☐	☐	☑	☐
Ease and ability to walk to gate from terminal	☑	☐	☐	☐
Please confirm longest walking distance		15 m		
State if track transit systems (TTS) are used	Yes ☐		No ☑	
State if passenger walkways are used	Yes ☐		No ☑	
Suitability and size of secondary airside retail	☐	☐	☐	☑
Effectiveness of immigration size and processes	☑	☐	☐	☐
Effectiveness of arrivals reclaim hall and BHS	☐	☐	☑	☐

15 Other relevant information

The increase of passenger turnover is 23.1% in comparison with year 2007.

Appendix 1: Airports own customer survey questionnaire.

Chişinău
AEROPORT

CHESTIONAR

Aeroportul Internaţional Chişinău tinde spre perfecţiune - nu în concepţia noastră, ci a Dvs. Vă rugăm să ne acordaţi cîteva minute, completînd chestionarul de mai jos şi transmiţîndu-l Biroului de informaţii. Opinia Dvs. exprimată va facilita îmbunătăţirea serviciilor noastre pînă la vizita Dvs. următoare.

Date personale

Ţara: ______________________ Oraşul: ______________________

Sexul Dvs. ☐ Masculin ☐ Feminin
Vîrsta Dvs. ☐ pînă la 25 ani ☐ 25-35 ani ☐ 35-45 ani ☐ 45-60 ani ☐ mai mult de 60 ani

Informaţii privind călătoriile Dvs.

De cîte ori, pe parcursul ultimului an, aţi beneficiat de serviciile Aeroportului Internaţional Chişinău?
☐ nici o dată ☐ 1-2 ☐ 3-5 ☐ 6-10 ☐ 11-20 ☐ mai mult de 20
În ce scopuri călătoriţi? ☐ Interes de serviciu ☐ Odihnă ☐ Altul
Călătoriţi deseori cu copiii? ☐ Da ☐ Nu
Cu ce mijloc de transport aţi călătorit ca să ajungeţi la aeroport? ☐ Transportul public ☐ Transportul personal

Evaluare aeroport

Curăţenia în aerogară şi pe teritoriul aferent - ☐ 1 ☐ 2 ☐ 3 ☐ 4 ☐ 5+
Funcţionalitatea şi confortul terminalului - ☐ 1 ☐ 2 ☐ 3 ☐ 4 ☐ 5+
Calitatea informaţiei auditive oferite - ☐ 1 ☐ 2 ☐ 3 ☐ 4 ☐ 5+
Calitatea informaţiei vizuale oferite (indicatoarele, panourile) - ☐ 1 ☐ 2 ☐ 3 ☐ 4 ☐ 5+
Activitatea biroului de informaţii şi administratorilor - ☐ 1 ☐ 2 ☐ 3 ☐ 4 ☐ 5+
Serviciile prestate în aeroport, în totalitate - ☐ 1 ☐ 2 ☐ 3 ☐ 4 ☐ 5+
Activitatea aeroportului, în general - ☐ 1 ☐ 2 ☐ 3 ☐ 4 ☐ 5+

Evaluarea personalului

Controlul vamal - ☐ 1 ☐ 2 ☐ 3 ☐ 4 ☐ 5 +
Controlul de frontieră - ☐ 1 ☐ 2 ☐ 3 ☐ 4 ☐ 5+
Controlul de securitate - ☐ 1 ☐ 2 ☐ 3 ☐ 4 ☐ 5+
Înregistrarea biletelor şi bagajului - ☐ 1 ☐ 2 ☐ 3 ☐ 4 ☐ 5+
Deservirea la vînzarea biletelor şi efectuarea altor operaţiuni de casă - ☐ 1 ☐ 2 ☐ 3 ☐ 4 ☐ 5+

Dvs. aţi beneficia de serviciile suplimentare, oferite contra plată, ca, de exemplu:

Transportarea bagajului la domiciliu/ la hotel ☐ da ☐ nu
Închirierea automobilului ☐ da ☐ nu
Rezervarea hotelului ☐ da ☐ nu
Punerea la dispoziţie a taxiului la aterizare ☐ da ☐ nu
Serviciile salonului de frumuseţe ☐ da ☐ nu
Sugestiile şi reclamaţiile Dvs. privitor la calitatea serviciilor aeroportuare

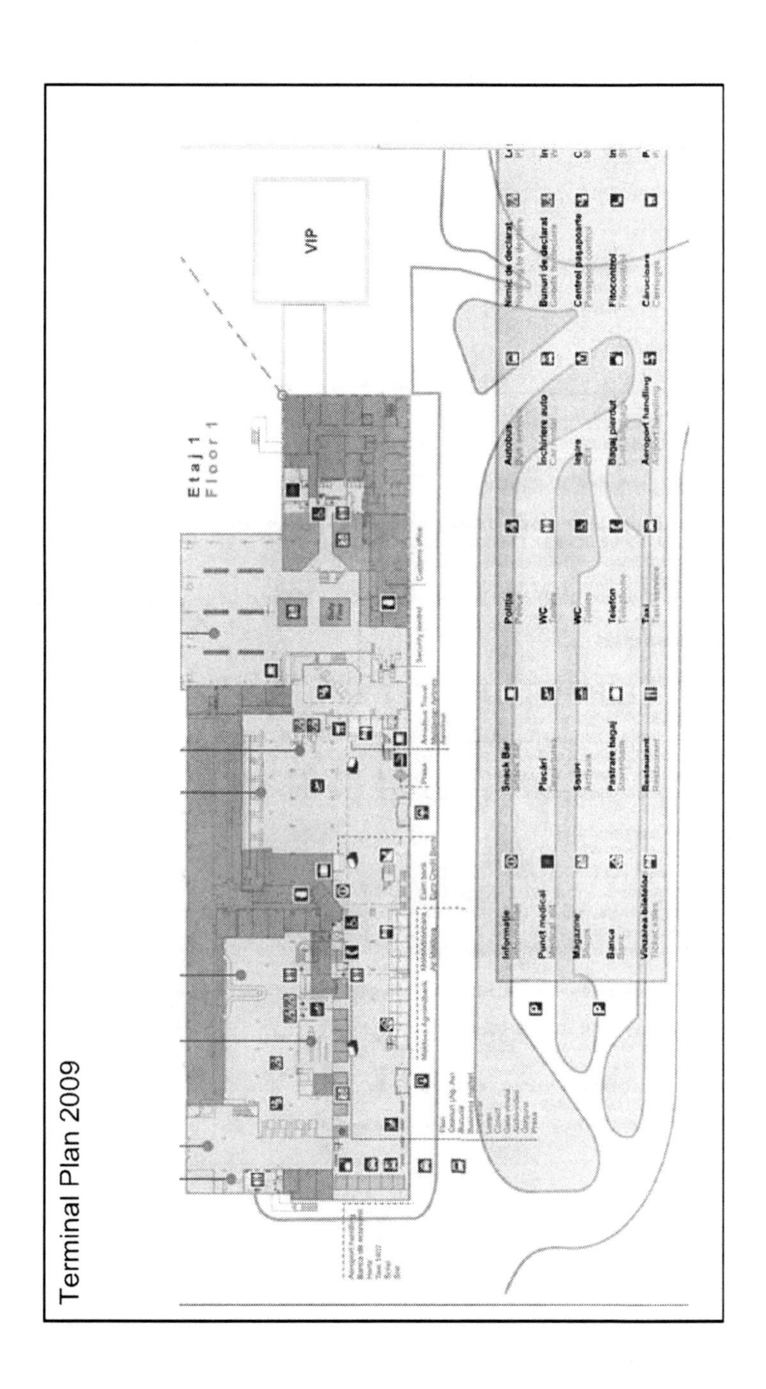
Terminal Plan 2009
Etaj 1
Floor 1
VIP
Informatie
Snack Bar
Politia
Autobus
Nimic de declarat
Punct medical
Plecari
WC
Inchiriere auto
Bunuri de declarat
Magazine
Sosiri
WC
Control pasaporte
Banca
Pastrare bagaj
Telefon
Bagaj pierdut
Fitocontrol
Vanzarea biletelor
Restaurant
Taxi
Aeroport handling
Carucioare

7.3 Airport research report: Cologne Bonn Airport, Germany

Airport name/country

Cologne Bonn Airport, Germany

Report by

Author, with supplement data collected by the author

Date

2009

1 Airport capacity/demand/miscellaneous issues

1.1 What is the current MPPA (2008)?

10.35 MPPA

1.2 What is the MPPA split between departing passengers and arriving passengers?

Departures	50%
Arrivals	50%
Total	100%

1.3 What is the forecasted MPPA?

2008, 10.35 MPPA 2009, N/A

1.4 What level of future infrastructure is planned in order to accomplish this growth?

Terminals 1 and 2 have a joint current capacity of 12 MPPA

1.5 Do you have published service level space standards (space per passenger)?

Not available

2 Check-in

Passenger operations

2.1 What is the technology split of check-in?

	Now	5 years' time
Conventional	95%	<20%
CUSS	0%	Unknown %
Branded self-service	4%	0%
Online	1% (BA)	Unknown %

2.2 What is the designed space per passenger in this zone (m^2/pax)?

Not available

Baggage operations

2.3 How many dedicated 'Baggage Drop' positions are used with CUSS/ SSCI?

N/A

2.4 How many oversized baggage drops are provided?

Three units are in operation now

2.5 Does the baggage handling system incorporate automatic or manual sortation?

T1 – Semi-automatic
T2 – Fully automatic belts

2.6 What is the floor area of the baggage handling system?

8000 m^2

3 Security

3.1 Do you have:

- **Centralised security only?**

Yes

- **Gate security only?**

Possible/safeguarded for design

- **Both centralised and gate security?**

Yes

3.2 Are security costs subsidised by the Government?

For passengers: Yes
For staff/flight crew: No

3.3 Do you process hand and hold baggage through central passenger security X-ray machines?

No – not permitted under current German legislation

3.4 What is the processing rate of passengers through the security functions?

Averages 120 per hour/X-ray and AMD unit

3.5 What is the designed space per passenger in this zone (m^2/pax)?

N/A

4 Departures immigration

4.1 Is there a requirement for departures immigration?

Yes – limited

4.2 On average how long does it take to process EU/non-EU passengers through the departures immigration process?

EU	N/A pax/min
Non-EU	N/A pax/min

4.3 What is the designed space per passenger in this zone (m^2/pax)?

N/A

5 Departures retail lounges

5.1 What % proportion of the terminal building area is dedicated to:

- **Airside retail?**

60% N/A m^2

- **Landside retail?**

40% N/A m^2

5.2 Is there retail within the piers/satellites?

No separate pier/satellite

6 Passenger delivery

6.1 What is the maximum airside walking distance witnessed for passengers?

150 m

6.2 What proportion (%) of passengers use disabled facilities (any special facilities)?

1500 pax/month use powered buggies or wheelchairs

6.3 What proportion (%) of stands/gates are pier/satellite served?

21 Contact stands 76 Remote stands

7 Arrivals immigration and transfers

7.1 How many arrivals immigration desks are used?

T1 = 14 T2 = 8

7.2 Typically how long does it take EU/non-EU passengers to progress through arrivals immigration?

EU N/A Minutes Non-EU N/A Minutes

7.3 Are biometric systems used/planned to be needed at immigration?

Not yet (staff will be required to use biometric facilities first)

7.4 Is there an airside transfer function and what percentages of arriving traffic are associated with transfers?

- **Airside transfer**

N/A % – mainly point-to-point traffic

- **Landside transfer**

N/A % – mainly point-to-point traffic

7.5 Is the airside transfer function only for full-service carriers or do low-cost carriers use them?

None – N/A

7.6 How many security search archways and screening machines are used?

T1 = 13
T2 = 9

7.7 What is the designed space per passenger in this zone (m^2/pax)?

N/A

8 Arrivals baggage reclaims

8.1 How many baggage reclaims are used?

T1 = 4
T2 = 4 (2 additional units are planned)

8.2 What is the designed space per passenger in this area (m^2/pax)?

N/A

8.3 Any special features?

N/A

9 Arrivals customs facilities

9.1 What is the area of the customs facilities?

N/A

9.2 What is the designed space per passenger in this area (m^2)?

N/A

10 Energy management systems

10.1 Are there energy management systems in place?

- **Lighting systems?**
- **Baggage systems?**
- **Passenger travelators (moving walkways)?**
- **Other?**

Energy management systems are in place to control:
Heating and cooling
Building management – system controls
Daily usage/consumption monitored

10.2 What is the energy consumption/generation strategy for the airport?

T2 – heating and cooling is via a conductive floor plate system

11 Waste management systems

11.1 What waste management strategies are in place?

Waste is recycled and broken down into three categories:
Paper
Glass
Biodegradable

11.2 Are there waste compactors on site?

There are compactors for each category of recycled waste

12 Fire strategy

12.1 How does the fire evacuation strategy for the main terminal building work?

There are:
- Smoke detection sensors
- Smoke exhausting vents
- Sprinklers
- T2 has 30 minute vaulted ceiling
- T2 has ducted fans system to expel smoke
- T2 has fire curtains

12.2 Do the building level changes experienced by the passengers create any problems for building evacuation?

No evacuation problems experienced to date

13 Apron questions

13.1 Could we have a layout of the airport/apron showing runways and taxiway positions?

Available ☑ See Appendix 2 (Terminals 1 and 2) Not available ☐

13.2 Explain the general functionality of the airfield.

Primary runway (14L/32R) serves 75% of movements
Secondary runway (25/07) serves 25% of movements
Other ....
There is a very small section of dual taxiways (A7) near the new Terminal 2 apron. This overcomes a congestion point for aircraft getting on to and off the 14L runway threshold

13.3 For what wing span/ICAO Code has the airfield been safeguarded?

Code E/F

13.4 What are the runway length(s)?

Runway (14L/32R)	length = 3850 m
Runway (25/07)	length = 2400 m
Runway (14R/32L)	length = 1863 m
Runway ( )	length = m

13.5 What is the runway operating mode (segregated/compass mixed mode, etc.)?

Simultaneous synchronised (mixed mode) use of 14L/32R and 25/07 cross runways mainly

13.6 What is the average and maximum taxi distance from runway to stand?

N/A

13.7 What is the average aircraft turn-around time?

Average 35 minutes

13.8 What is the passenger apron size?

See layout drawings provided

13.9 What is the cargo throughput (tonnes/annum)?

570 000–620 000 tonnes/annum

13.10 What is the cargo shed size?

60 000 m^2

Appendix 2: Internal photographs of Terminals 1 and 2

Terminal 2 Walk-through check-in

Terminal 1 Branded self-service check-in

Terminal 1 Euro satellite – low-cost carrier satellite

7.4 Airport research report: Adelaide Airport, Australia

Airport name/country

Adelaide Airport – Australia

Report by

Vince Scanlon

Date

May 2009

1 Airport capacity/demand/miscellaneous issues

1.1 What is the current MPPA (2009/10)?

7 million

1.2 What is the MPPA split between departing passengers and arriving passengers?

Departures	50%
Arrivals	50%
Total	100%

1.3 What is the forecasted MPPA for the next 5 years?

2010, 7.03M 2011, 7.24M 2012, 7.63M 2013, 8.03M 2014, 8.35M

1.4 What level of future infrastructure is planned in order to accomplish this growth?

Main terminal expansion
Terminal 2 new build for regional traffic
Multilevel carpark

1.5 Do you have published service level space standards (space per passenger)?

Yes

2 Check-in

Passenger operations

2.1 What is the technology split of check-in?

	Now	5 years' time?
Conventional	42 desks	42 desks
CUSS	0 kiosks	12 kiosks
Branded self-service	14 kiosks	0 kiosks
On-line	20% pax	40 % pax

2.2 What is the designed space per passenger in this zone (m^2/pax)?

1.5 m^2 per pax
1.5 m^2 per visitor

Baggage operations

2.3 How many dedicated 'Baggage Drop' positions are used with common user self-service (CUSS)/self-service check-in (SSCI)?

Varies due to common use counter allocation

2.4 How many oversized baggage drops are provided?

Two

2.5 Does the baggage handling system incorporate automatic or manual baggage sortation? Is there an airline baggage sortation technology preference?

Automatic sortation to individual airline allocated make-up laterals

2.6 What is the floor area of the baggage handling system?

Domestic, 3600 m^2
(International, 1100 m^2)

3 Security

3.1 Do you have:

- **Centralised security only?**

Yes

- **Gate security only?**

No

- **Both centralised and gate security?**

No

3.2 Are security costs subsidised by the Government?

No

3.3 Do you process hand and hold baggage through central passenger security X-rays machines?

No – hold luggage is via CBS X-ray units in line with the BHS

3.4 What is the processing rate of passengers through the security functions?

Up to 300 pax per hour (LAGs separate) in peak

3.5 What is the designed space per passenger in this zone (m^2/pax)?

1.0 m^2

4 Departures immigration

4.1 Is there a requirement for departures immigration?

Yes

4.2 On average how long does it take to process EU/non-EU passengers through the departures immigration process?

By Australian Customs		
American	pax/min	All approximately 1 minute
Non-American	pax/min	
EU	pax/min	
Non-EU	pax/min	

4.3 What is the designed space per passenger in this zone (m^2/pax)?

1.0 m^2

5 Departures retail lounges

5.1 What % proportion of the terminal building area is dedicated to:

- **Airside retail?**

  N/A m^2 12%

- **Landside retail?**

  N/A m^2 Less than 1%

5.2 Is there retail within the piers/satellites?

Minimal

6 Passenger delivery

6.1 What is the maximum airside walking distance witnessed for passengers?

Approximately 720 m

6.2 What proportion (%) of passengers use disabled facilities (any special facilities)?

Approximately 5% (deafness-friendly telephones and hearing loops at check-in)

6.3 What proportion (%) of stands/gates are pier/satellite served?

100%

6.4 Is there a tracked transit system (mass people movement system)? If yes, what is its function – (landside terminal transfers or airside terminal to pier connections)?

No

7 Arrivals immigration and transfers

7.1 How many arrivals immigration desks are used?

12

7.2 Typically how long does it take American/non-American/EU/non-EU passengers to progress through arrivals immigration?

By Australian Customs

American	min	All approximately 1 minute
Non-American	min	
EU	min	
Non-EU	min	

7.3 Are biometric systems used/planned to be needed at immigration?

Customs Smartgate – Australian and New Zealand citizens only at this stage

7.3.1 How long does it take to process passengers using this technology?

Australia/New Zealand only

American	min	Approximately 45 seconds
Non-American	min	
EU	min	
Non-EU	min	

7.4 Is there an airside transfer function and what percentages of arriving traffic are associated with transfers?

- **Airside transfer**

%	No

- **Landside transfer**

%	No

7.5 Is the airside transfer function only for full-service carriers or do low-cost carriers use them?

N/A

7.6 How many security search archways and screening machines are used?

None used for arrivals – passengers segregated from departures concourse

7.7 What is the designed space per passenger in this zone (m^2/pax)?

1.5 m^2

8 Arrivals baggage reclaims

8.1 How many baggage reclaims are used?

4 domestic / 1 international

8.2 What is the designed space per passenger in this area (m^2/pax)?

1.5 m^2

8.3 Any special features?

No

9 Arrivals customs facilities

9.1 What is the area of the customs facilities?

525 m^2

9.2 What is the designed space per passenger in this area (m^2)?

$1.5\,m^2$

10 Energy management systems

10.1 Are there energy management systems in place?

- **Lighting systems?**
- **Baggage systems?**
- **Passenger travelators (moving walkways)?**

Other?

Back-up generator for all facilities including light, power, BHS, emergency lights, aerobridge, escalators, travelators, check-in

10.2 What is the energy consumption/generation strategy for the airport?

Mains power 3500 kW with 2×1340 kW back-up generators

11 Waste management systems

11.1 What waste management strategies are in place? Is waste broken down into categories?

Waste is separated into general waste and recycle (glass/plastic) products via compactors

11.2 Are there waste compactors on site?

Yes – 2 as per 11.1

12 Fire strategy

12.1 How does the fire evacuation strategy for the main terminal building work?

Fire alarm in activated zone will automatically time out to evacuation (if no intervention) and adjacent zones will cascade into alert status

12.2 Do the building level changes experienced by the passengers create any problems for building evacuation?

No – building evacuation zones are separated horizontally throughout the building

13 Apron questions

13.1 Please attach a layout of the airport/apron showing runways and taxiway positions.

Available ☑ Not available □
See Appendix 3

13.2 Explain the general functionality of the airfield.

Primary runway (05/23) serves up to Code F aircraft and is primary runway
Secondary runway (12/30) serves up to Code C aircraft and is secondary runway in adverse wind conditions
Other ....

13.3 For what wing span/ICAO Code has the airfield been safeguarded?

Generally Code E 47.5 m plus dispensation for Code F operation on specified route

13.4 What are the runway length(s)/orientations?

Runway (05/23) length = 3100 m
Runway (12/30) length = 1650 m
Runway ( ) length = m
Runway ( ) length = m

13.5 What is the runway operating mode (segregated/compass mixed mode, etc.)?

Compass mixed

13.6 What is the average and maximum taxi distance from the runway to stand?

Average 1425 m

13.7 What is the average aircraft turn-around time?

Low-cost carrier	30	mins (B737-800 say)
Full-service carrier	40	mins (B737-800 say)

13.7.1 How is the passenger boarding managed for low-cost flights?

All aircraft board via passenger boarding bridges (PBBs). Low-cost airlines also board via rear stair (apron) depending upon weather

13.8 What is the passenger apron size?

85 370 m^2

13.9 What is the cargo throughput (tonnes/annum)?

8582 tonnes export
9090 tonnes import

13.10 What is the cargo shed size?

2 main handlers
Australian Air Express 4780 m^2
Toll 3100 m^2

13.11 What is the cargo apron (stand) area?

Single bay 6970 m^2

13.12 Are the low-cost stands served with airbridges?

Yes – refer to 13.7.1

14 Passenger experience

14.1 Are there any customer satisfaction surveys carried out across the airport? If 'yes', could we have a copy of this material?

Yes – ACI service quality survey

14.2 For each statement listed below please tick one box that best describes the effectiveness of the airport function when viewed from a passenger's perspective.

	Very good	Good	Average	Poor
Ease of entering the airport complex by train	☐	☐	☐	☑ N/A
Ease of entering the airport complex by car	☐	☑	☐	☐
Ease of entering the airport complex by public bus	☐	☑	☐	☐
Effectiveness of check-in processes	☑	☐	☐	☐
Suitability and size of *landside* retail offer	☐	☐	☑	☐
Speed and effectiveness of security processes	☐	☑	☐	☐
Suitability and size of *airside* primary retail offer	☐	☑	☐	☐
Ease and ability to walk to gate from terminal	☐	☑	☐	☐
Please confirm longest walking distance	750 m			
State if TTS are used	Yes ☐ No ☑			
State if passenger walkways are used	Yes ☑ No ☐			
Suitability and size of secondary airside retail	☐	☑	☐	☐
Effectiveness of immigration size and processes	☐	☑	☐	☐
Effectiveness of arrivals reclaim hall and BHS	☐	☑	☐	☐

15 Other relevant information

Adelaide Airport is a common use combined domestic and international terminal

Appendix 3: Terminal 1 and aerodrome plan

Adelaide Airport Apron Terminal 1

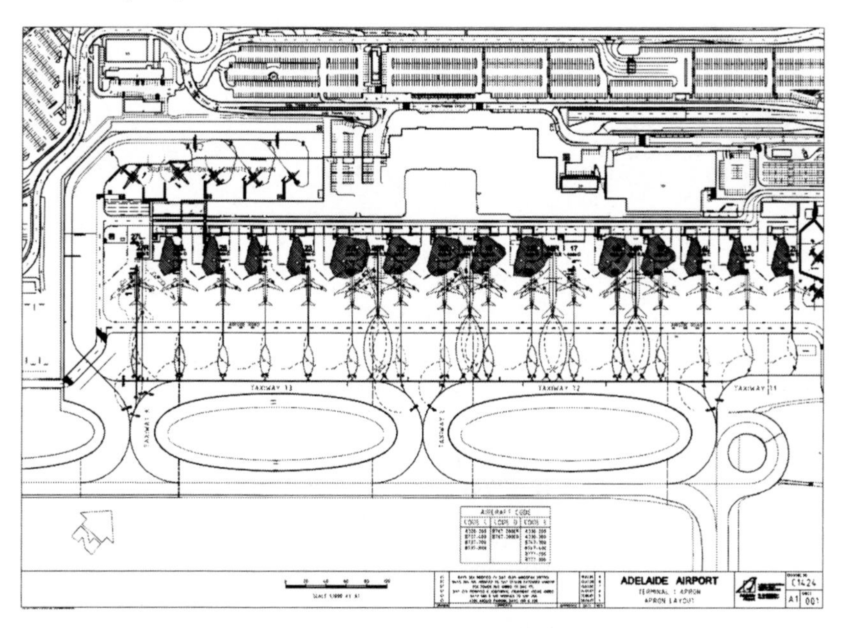

Aerodrome Plan Open V Runway

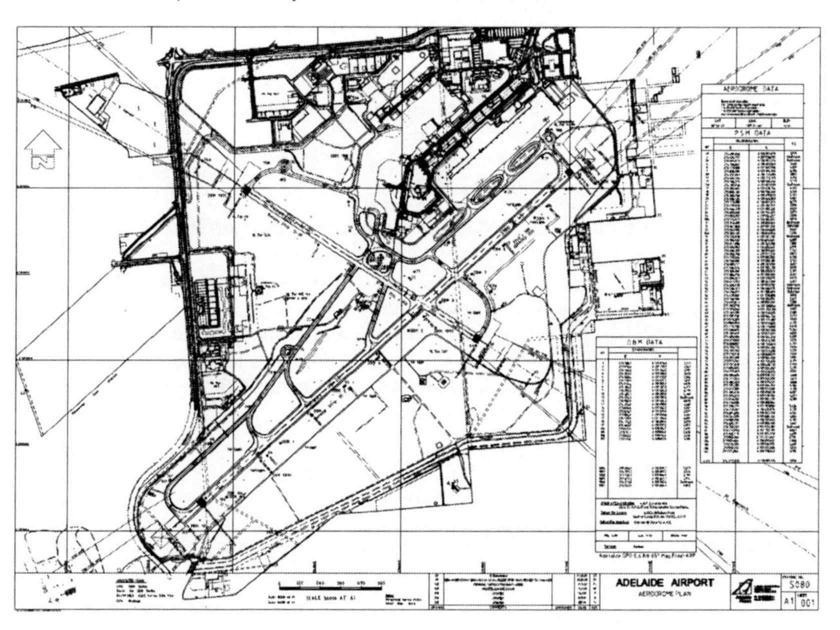

7.5 Airport research report: low-cost bespoke terminal, Luton Airport, UK

Airport name/country

Luton Airport, UK

Report By

Author

Date

2009

Airport parameters	Quantity/units/description
Terminal data	
Number of terminals	2 terminals 30 000 m^2 in total T2 = 20 000 m^2 as a dual level building The new second terminal was opened in Autumn 1999 with 60 check-in desks Terminal is fitted with pay on entry executive lounge
% LCC airlines using the facility	90% 10.2 million passengers used the airport in 2008 95% of passengers fly on scheduled services 5% fly on charter services 87% of passengers were on international flights, 13% of passengers were on domestic flights Low cost = 85% Full service = 10% Charter = 5%
Number of check-in desks	60 desks with full common user terminal equipment (CUTE) capability 20% of self-service airport owned check-in went operational 2005 Baggage handling system is a sophisticated tub tray system – unusual in that the tubs are used from a check-in collector throughout the system but is effective
Number of reclaim units	4 reclaim units
Number of security search units	6 search units
Number of piers	Most recent pier went operational in October 2005

Table (cont.) Airport parameters	Quantity/units/description
Airfield data	
Number of aircraft stands Code C	10
Number of aircraft stands Code D and above	9
Passenger numbers/ growth (MPPA)	1999/2000 5.7 MPPA 2003/04 10.1 MPPA 2009/10 14.58 MPPA 2017/8 20.07 MPPA
Number of runways	1 (~2500 m)
Capital cost/year	Cost £40M (T2) in 1999, Capacity = 10 MPPA

7.6 Airport research report: low-cost terminal, Frankfurt Hahn Airport, Germany

Airport name/country

Frankfurt Hahn Airport, Germany

Report by

Author

Date

2009

Airport parameters	Quantity/units/description
Terminal data	
Number of terminals	2 terminals estimated 20 500 m^2 in total
% LCC airlines using the facility	95%
Number of check-in desks	17 desks
Number of reclaim units	4
Number of security search units	2 known – 4 maximum
Number of piers	1 (contact stand)
Airfield Data	
Number of aircraft stands Code C	6 estimated
Number of aircraft stands Code D and above	3 estimated
Passenger numbers/ growth (MPPA)	2005 3.079 MPPA 2006 3.705 MPPA 2007 4.015 MPPA 2008 3.939 MPPA 2009 Not available
Number of runways	1 runway 3045 m, currently proposing to extend to 3800 m Expansion of length of runway to permit more cargo operations
Capital cost/year	T1 (10 000 m^2) in early 2001 expansion and refurbishment cost = £8.2M (DEM11.5M). Ryanair on board late 2001 T2 built (~10 500 m^2) in 2003 = £3.57M (5M euros) Capacity claimed to be '4–5 MPPA'.
Further development factors	Frankfurt Hahn have agreed capital injection of 42M euros (£30M) to be invested between 2005 and 2009

Index

Lightning Source UK Ltd.
Milton Keynes UK
UKOW031113250412

191426UK00002B/16/P